Sensors Update

Volume 10

Sensor Technology –
Applications – Markets

Related Wiley-VCH titles:

Sensors Vol. 1–9
edited by W. Göpel, J. Hesse, J. N. Zemel

ISBN 3-527-26538-4

Sensors Applications
edited by J. Hesse, J. W. Gardner, W. Göpel

Vol. 1

Sensors in Manufacturing
edited by H. K. Tönshoff, I. Inasaki

ISBN 3-527-29558-5

Vol. 2

Sensors in Intelligent Buildings
edited by O. Gassmann, H. Meixner

ISBN 3-527-29557-7

Upcoming Volumes:

Sensors in Medicine and Health Care
Sensors in Household Appliances
Sensors in Aerospace Technology
Sensors in Environmental Technology
Sensors in Automotive Technology

Sensors Update

Edited by
H. Baltes, G. K. Fedder, J. G. Korvink

Volume 10

Series Editors

Prof. Dr. H. Baltes
Physical Electronics Laboratory
ETH Hoenggerberg, HPT-H6
8093 Zürich
Switzerland

Prof. Gary K. Fedder
ECE Department &
The Robotics Institute
Carnegie Mellon University
Pittsburgh, PA 15213-3890
USA

Prof. Dr. Jan G. Korvink
Institute for Microsystem
Technology
University of Freiburg
Georges-Köhler-Allee 103
79110 Freiburg
Germany

The series Sensors Update was founded by
Prof. Dr. H. Baltes, Prof. Dr. W. Göpel and Prof. Dr. J. Hesse.

Library of Congress Card No.: applied for

British Library Cataloguing-in-Publication Data:
A catalogue record for this book is available from the British Library

Die Deutsche Bibliothek – CIP-Cataloguing-in-Publication Data
A catalogue record for this publication
is available from Die Deutsche Bibliothek

ISBN 3-527-30361-8

Composition: K+V Fotosatz GmbH, Beerfelden
Printing: betz-druck GmbH, Darmstadt
Bookbinding: J. Schäffer GmbH & Co. KG, Grünstadt
Printed in the Federal Republic of Germany

Preface

The rapid development of sensors with computer-compatible output signals has been quite remarkable since its beginnings in the mid 1970s. In addition to the technological advances driving the sensor industry, there has been a boom in new applications and market sectors where new sensor technologies can be applied, thus creating an increasing demand for sensors and associated instrumentation through the turn of the century. In turn, research efforts in the sensor field are growing even more, as is apparent from the extent and number of old and new conferences, journals, and exhibitions as well as new companies. In view of the diversity of sensor technologies and applications, to stay informed is a big challenge. This is where *Sensors Update* plays a unique role by providing timely and critical reviews.

The aim of *Sensors Update* is to help the reader stay at the cutting edge of the field. Built upon the 9-volume series *Sensors*, *Sensors Update* with this 10th volume has now surpassed the size of its parent publication and has also become a journal. *Sensors Update* presents an overview of highlights in the field on a regular basis. Coverage includes current developments in materials, design, production and applications of sensors, signal detection and processing, as well as new sensing principles. The sensor market as well as important aspects such as standards or patents are also addressed.

'Sensor Technology' reviews highlights in applied and basic research, 'Sensor Applications' covers new or improved applications of sensors, and 'Sensor Market' provides an overview of suppliers, market trends or patent information for a particular sector.

Timely reviews are highly desirable, but are a lot of work to write and edit. Thus, the series editors wish to thank their colleagues who have contributed to this important enterprise by writing, reviewing, or editing articles. Thanks are also due to the publishers, Wiley-VCH and their staff Dr. Peter Gregory, Dr. Claudia Barzen, Hans-Jochen Schmitt and Dr. Jörn Ritterbusch for their support.

With the present volume 10, a new co-editor, Prof. Gary K. Fedder, is replacing Prof. Joachim Hesse who has retired from this series after over a decade of co-editing *Sensors* and *Sensors Update*. Gary Fedder is associate professor of electrical and computer engineering and robotics at Carnegie Mellon University, Pittsburgh, PA, USA. More information about his work can be found at www.ece.cmu.edu/~fedder.

H. Baltes, Zürich   G. K. Fedder, Pittsburgh   J. G. Korvink, Freiburg

Editorial Advisory Board

Contents

List of Contributors

Hassan Y. Aboul-Enein
Pharmaceutical Analysis Laboratory
Biological & Medical Research Dept.,
MBC-03
King Faisal Specialist Hospital & Research
Centre
P.O. Box 3354
Riyadh 11211
Saudi Arabia

Karl F. Böhringer
Department of Electrical Engineering
University of Washington
Box 352500
Seattle WA 98195-2500
USA

Karl S. Booksh
Department of Chemistry and Biochemistry
Arizona State University
Tempe, AZ 85287
USA

Teruo Fujii
Underwater Technology Research Center
Institute of Industrial Science
University of Tokyo
4-6-1 Komaba, Meguro-ku
Tokyo 153-8505
Japan

Sarah Garrod
Department of Electrical Engineering
University of Washington
Box 352500
Seattle, WA 98195-2500
USA

Yael Hanein
Department of Electrical Engineering
University of Washington
EE/CSE Bldg., M242
Box 352500
Seattle, WA 98195-2500
USA

Sean Hoyt
Department of Electrical Engineering
University of Washington
Box 352500
Seattle, WA 98195-2500
USA

Sam McKennoch
Department of Electrical Engineering
University of Washington
Box 352500
Seattle, WA 98195-2500
USA

Thomas Laurell
Department of Electrical Measurements
Lund Institute of Technology
Lund University
P.O. Box 118
22100 Lund
Sweden

Andrey Legin
Laboratory of Chemical Sensors
Chemistry Department
St. Petersburg University
St. Petersburg
Russia

Johanngeorg Otto
Fachhochschule Aalen
Beethovenstr. 1
73428 Aalen
Germany

William R. Penrose
Illinois Institute of Technology
Department of Biological, Chemical and
Physical Sciences
LS-178
3101 South Dearborn Street
Chicago, IL 60616-3793
USA

Carole Rossi
LAAS-CNRS
7 ave du colonel Roche
31077 Toulouse cedex 4
France

Alisa Rudnitskaya
Laboratory of Chemical Sensors
Chemistry Department
St. Petersburg University
St. Petersburg
Russia

Raluca-Ioana Stefan
Department of Chemistry
University of Pretoria
0002 Pretoria
South Africa

Joseph R. Stetter
Department of Biological, Chemical and
Physical Sciences
148a Life Sciences Building
Illinois Institute of Technology
3300 South Federal Street
Chicago, IL 60616-3793
USA

Atsuko Takamatsu
PRESTO
Japan Science and Technology Corporation
(JST)
Saitama
Japan

Jakobus F. van Staaden
Department of Chemistry
University of Pretoria
0002 Pretoria
South Africa

Yuri Vlasov
Laboratory of Chemical Sensors
Chemistry Department
St. Petersburg University
St. Petersburg
Russia

Joseph Wang
Department of Chemistry
and Biochemistry
New Mexico State University
Las Cruces, NM 88003
USA

A. O. Dennis Willows
Department of Zoology
University of Washington
Seattle, WA 98195
USA

Denise M. Wilson
Department of Electrical Engineering
University of Washington
Box 352500
EE/CSE Bldg., M222
Seattle, WA 98195-2500
USA

Russell C. Wyeth
Department of Zoology
University of Washington
Seattle, WA 98195
USA

PART 1

Sensor Technology

1.1 Biocatalytic Porous Silicon Microreactors

T. LAURELL,
Lund Institute of Technology, Lund University, Lund, Sweden

Abstract

This review covers the development of the porous silicon micro immobilized enzyme reactor (PS µIMER). The efforts in tailoring the pore morphology to provide an optimal highly catalytic microfluidic component for integration in chemical microanalysis systems are reviewed. Optimization was performed with respect to silicon dopant type, thickness of the porous silicon layer, and the anodization conditions. Applications of the µIMER to sugar monitoring and to glutamate and protein analysis are described.

Keywords: Porous silicon; enzyme; micro total analysis system; microreactor; biosensor; proteomics

Contents

1.1.1   Introduction

The rapid development of miniaturization and the drive towards chip-based analysis systems in analytical chemistry puts demands on new approaches to integrate macro-scale bioanalytical functions. This development is partly fuelled by new advances in both microstructure technology and materials science, which now permits the design of complex microsystems comprising, eg, actuation devices for microfluidic sample transport, mixing structures, valves and detecting units. The concept of the micro total analysis (μTAS) system was first described by Manz et al. in 1990 [1] and has now developed into a highly interdisciplinary research field with a broad international research community, in which the advances of μTAS developments to a large extent are reported at the μTAS meetings [2–5].

To implement biospecific detection and monitoring in chemical microsystems, enzymes are commonly used in conjunction with a suitable detection technique, eg, electrochemical, optical, conductometric, calorimetric, etc. The traditional way of maximizing the readout of the analyte is to pass the sample through a packed bed of porous beads with immobilized enzymes to ensure a high catalytic conversion and thus generate a high concentration of products that subsequently can be monitored. Although monocrystalline silicon wafers in principle display the same surface chemistry (SiO_2) as controlled pore glass, and thus is easy to surface modify using the same procedures as described for glass beads [6], it is not necessarily an optimal surface for implementing biospecificity in μTAS owing to its extreme flatness and low surface area properties. Flow lines and interconnections in μTAS, fabricated by means of conventional anisotropic, isotropic, or dry etching techniques are defined by very flat surfaces and are thus not very well suited for high-density immobilization of bioactive molecules such as enzymes. In contrast, the surface-enlarging properties of porous silicon (PS) offer exciting possibilities for the manufacture of integrated highly active chemical functionalities in a microformat at the chip level.

In this review, strategies to increase catalytic performance and aspects of the design of chip-integrated microreactors are discussed and applications of PS enzyme reactors are presented.

1.1.2   Silicon-Integrated Microreactors

Early work on chip-integrated micro enzyme reactors comprised an anisotropically etched meander-shaped single V-groove channel in $\langle 100 \rangle$-silicon, in which
glucose oxidase was immobilized on the silicon surface [7, 8]. The reactor covered a chip area of 19×28 mm. Although the surface area of the reactor was
moderate, the sensing principle was well demonstrated by means of electrochemical detection of the H_2O_2 produced in the catalytic reaction. With respect to
problems of the pressure drop in a single-channel reactor, an improved design
was demonstrated by Xie et al., who proposed a parallel V-groove reactor structure, 5×1 mm [9]. To obtain a biosensing component, the channel array was activated with enzyme, realizing sensor concepts for, eg, penicillin and hydrogen
peroxide. The sensing mechanism in this application monitored the temperature
rise in the sample plug caused by the exothermic enzymatic reaction [10]. In
1994, Strike et al. [11] also demonstrated a V-groove-type micro immobilized enzyme reactor (μIMER) for glucose monitoring similar to that described by Suda
and co-workers [7, 8].

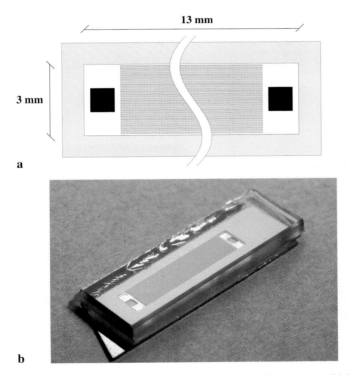

Figure 1.1-1. (**a**) Schematic diagram of the trench channel microreactor. (**b**) Microreactor
structure with anodically bonded glass lid.

An optional approach to the V-groove microreactor design was proposed in 1994 by Laurell and Rosengren [12], where the microreactor structure was defined by a parallel channel array based on 30 vertical trenches, 50 µm wide and 250 µm deep, covering a chip area of 3×13 mm (Figure 1.1-1 a). The vertical channels were etched by anisotropic etching of $\langle 110 \rangle$ silicon. The microreactor was sealed by anodic bonding of a glass lid to the chip surface and flow interconnections were defined by holes from the rear of the chip (Figure 1.1-1 b).

The fundamental strategy of the deep channel design was to obtain a larger surface area and thus to achieve a microstructure with a higher enzyme activity per mm^2 of silicon. It is evident that the deep channel design offers a higher surface area than the V-groove design per unit chip area.

Figure 1.1-2. Inlet of the two µIMERs with either **(a)** 50 µm or **(b)** 20 µm channel width.

Glucose oxidase was used throughout the continuing work when recording the enzyme activities for different microreactor designs. For a more detailed description of the enzyme activity monitoring protocol used, see [13]. The need for greater enzyme activity was clearly seen in the early applications of the μIMER [12] and thus the need for greater surface area was clearly expressed. An evident way to increase the surface area of the μIMER was to increase the aspect ratio of the channels, which is proportional to the ratio of channel height to channel width. The effect of increasing the channel density was demonstrated in [14], where μIMERs with channel dimensions of 50×250 μm and 30 channels in parallel were compared with 20×250 μm and 75 channels in parallel. Both μIMERs occupied a chip area of 3×13 mm but offered surface areas of 178 and 515 mm², respectively. Figure 1.1-2 shows a low-angle view of the inlet of both μIMERs.

The enzymatic turnover of the two reactor geometries is shown in Figure 1.1-3. The two plots for the 20 μm channel reactor show a comparison of the enzyme activities obtained from reactors either provided with a thick oxide layer prior to the enzyme immobilization or with only the native oxide. As seen in the plots, the enzyme activities were not influenced by the density of the oxide layer and hence subsequent generations of reactors used only the native oxide layer. It was also found that the estimated maximum substrate turnover rate, V_{max}, for the two microreactor designs differed in proportion to their internal surface area (Table 1.1-1).

In comparison with the surface area that a corresponding controlled pore glass bead microreactor volume would provide, an estimate indicated that the microreactor surface area had to be increased by a factor of 300 in the chip-based μIMER. This could be done either by increasing the channel density, width ≤/μmg, and mak-

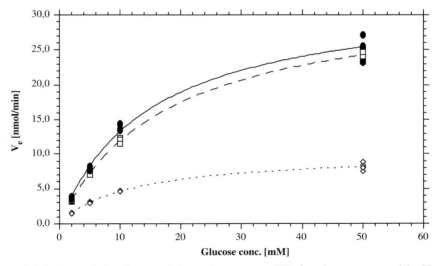

Figure 1.1-3. Recorded substrate (glucose) turnover (V_e) for the reactors with 50 and 20 μm wide channels. – • – 20 μm channel, native oxide; – □ – 20 μm channel, thick oxide; – ◇ – 50 μm channel, native oxide.

Table 1.1-1. Comparison of the catalytic turnover for the µIMERs with 20 and 50 µm channel widths, showing their corresponding maximum turnover rate, V_{max}, and apparent Michaelis constant, K_{mapp}

IMER	Surface area (mm^2)	V_{max} (nmol/min)	$K_{m\,app}$ (µmol)
50 µm channel IMER	178	10	12
20 µm channel IMER	515	~36	~17
Oxidized 20 µm channel IMER	515	~34	~19

ing deeper channels, > 300 µm, or by introducing a surface-enlarging matrix in the reactor. From the microprocess technology point of view, high aspect ratio channel networks, < 1 µm wide, 300 µm deep and 10 mm long, are not easily produced in large numbers at low prices. Hence other ways of increasing the surface area in the microreactors were therefore sought and the idea of using porous silicon as a carrier matrix for immobilized enzymes in chip-integrated microreactors was introduced in 1995 [14].

1.1.3   Porous Silicon –
An Inherent Surface-Enlarging Matrix

1.1.3.1   Background

The formation of porous silicon was first reported by Uhlir in 1956 [15] and initially the material found potential applications as an insulating material in integrated circuit design [16, 17]. In 1990 the active optical properties of PS were discovered [18] and hence research on the electroluminescence of porous silicon increased dramatically. The first applications to PS in chemistry were demonstrated by Smith and Collins, using silicon chip-integrated PS membranes as porous electrochemical contacts for an Ag/AgCl reference electrode on an electrochemical sensor chip [19].

The direct utilization of PS as a chemically enhancing material was not initiated until the mid-1990s, when the surface-enlarging properties of PS as a carrier matrix for an immobilized enzyme (glucose oxidase) was investigated [14, 20]. Schoning and co-workers later demonstrated the surface-enlarging benefits of PS in an electrolyte insulator semiconductor (EIS) component, monitoring the capacitance change of the EIS as a function of the pH change induced by the enzymatic conversion of penicillin. The device was developed as a biosensor for penicillin as penicillinase was adsorbed on the porous sensor surface [21, 22].

1.1.3.2   PS Etching

Porous silicon is formed in an electrochemical anodization process in which hydrofluoric acid is mixed with a surfactant, commonly ethanol, to lower the surface tension of the electrolyte. The formation of individual pores is initiated as holes migrate through the bulk silicon crystal, guided by the applied electrostatic field, and reach the electrolyte interface where a silicon fluoride complex is formed and dissolved in the electrolyte. A more thorough description and analysis of the pore formation process can be found in [23]. PS displays a wide range of porosities and pore sizes. The pore morphology obtained is dependent of a wide range of process conditions, some of the most important of which are as follows:

- silicon substrate dopant type and concentration;
- illumination conditions;
- anodization conditions: current density;
- crystal orientation;
- surface tension properties of the anodization electrolyte.

In order to find a porosity that is optimal as an enzyme-activated catalytic surface, an optimization process has to be carried out with respect to the process parameters that influence the pore formation.

A schematic diagram of a typical anodization cell for porous silicon etching is shown in Figure 1.1-4. To withstand the harsh electrolyte conditions, the etch cell should be made of an HF-resistant material. The cell developed by our group was made of poly(vinyl difluoride) (PVDF). The two electrode grids were made of a fine-meshed platinum grid and the optically transparent window was initially made of Nalgene® (Nunc, Denmark). As Nalgene® with time changed its transparency owing to the repeated exposure to HF, the material of the optical window was changed to sapphire. The different cell compartments were sealed with acid-resistant O-rings. A constant-current source was connected to the two Pt electrodes. When etching p-type silicon, the backside of the wafer was doped with boron to provide a low ohmic contact with the electrolyte of the anodic cell side.

In order to generate spatially defined components based on PS, the anodization has to be masked to predefined regions. This can be done by employing an HF-resistant mask and typically polymer films, silicon nitride, metal films, and highly doped n-type silicon patterning can be used [24, 25].

An attractive feature of PS when using the material as a carrier matrix for biochemical monitoring is the simple procedure for immobilizing enzymes, antibodies, or other ligands. As PS is generated and subsequently washed in deionized water, the surface of the porous matrix is spontaneously oxidized, leaving a silicon dioxide layer as the coupling surface. Figure 1.1-5 shows schematically the steps involved in a standard enzyme immobilization procedure [6].

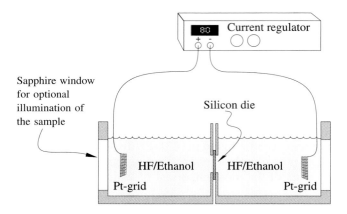

Figure 1.1-4. Electrochemical cell for porous silicon etching.

1.1.4   PS as a Carrier Matrix for Immobilized Glucose Oxidase on Planar Substrates

The initial work on enzyme-activated PS that was reported by our group was performed on planar PS silicon substrates [20]. n-Type (phosphorus, 0.1 Ω cm) silicon samples were anodized at three current densities (10, 50, and 100 mA/cm^2) to obtain samples with various porous properties. The electrolyte composition was (48%)HF–(96%)ethanol (1:1). The samples were washed and stored in deionized water. After immobilizing glucose oxidase on the porous matrix according to the scheme in Figure 1.1-5, the catalytic activity of the different surfaces achieved was monitored using a colorimetric assay with Trinder reagent [26]. The results are displayed in Figure 1.1-6, showing a 33-fold increase in the catalytic turnover for the sample anodized at 10 mA/cm^2 as compared with a corresponding nonporous reference sample. The sample surfaces were investigated by scanning electron microscopy (SEM) and, as expected, the low current density sample displayed a finer porous morphology (Figure 1.1-7 a–c) and hence a larger surface area for enzyme immobilization.

1.1.5   PS in the Trench Channel µIMER

The findings of increased catalytic turnover using porous silicon as the enzyme carrier matrix, triggered further work to combine PS with the trench channel µIMER and utilize the surface-enlarging properties of both the high aspect ratio parallel channel structure and the porous matrix. µIMERs with 30 parallel chan-

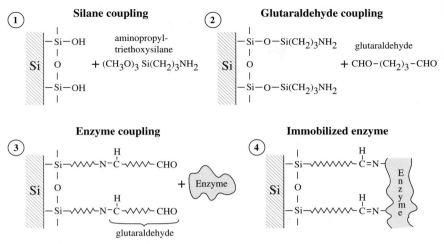

Figure 1.1-5. Principal steps, 1–4, in the immobilization of an enzyme on a silicon dioxide surface.

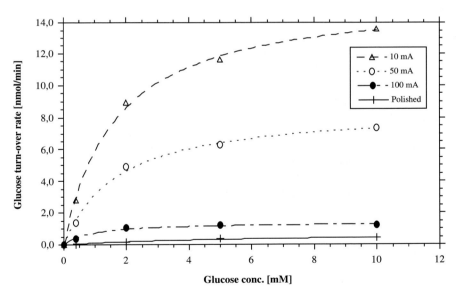

Figure 1.1-6. Catalytic turnover for the three PS samples anodized at 10, 50 and 100 mA/cm^2.

nels, channel/wall width of 50/50 μm, and a depth of 250 μm were anodized at 10, 50, and 100 mA/cm^2. Figure 1.1-8 a–c display the cross-sections of the three porosified μIMERs. The anodization current density was calculated with respect to the true surface area of the chip, i.e., including the increased surface area from the deep parallel channel structure.

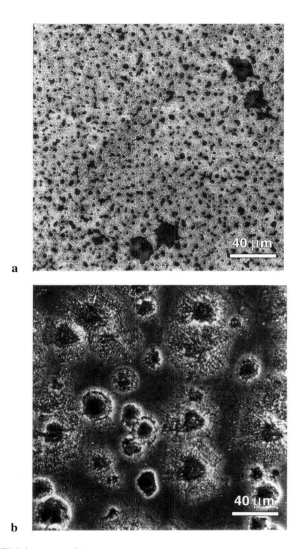

Figure 1.1-7. SEM images of the three glucose oxidase-activated planar PS surfaces: (**a**) 10, (**b**) 50.

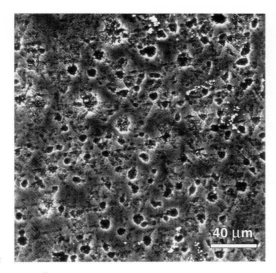

c

Figure 1.1-7 c. 100 mA/cm^2.

a

Figure 1.1-8. Cross-section SEMs of the PS μIMERs anodized at (**a**) 10.

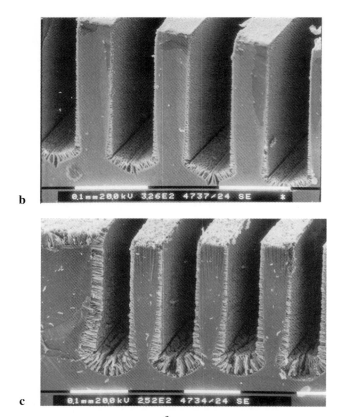

Figure 1.1-8. (**b**) 50, and (**c**) 100 mA/cm^2.

1.1.6   Glucose Monitoring

After anodization, the µIMERs were activated with glucose oxidase and the enzyme kinetic data (Figure 1.1-9) were derived according to the procedures described above. The increase in turnover rate for the best performing reactor, anodized at 50 mA/cm^2, was found to be 170 times higher than for the nonporous reference µIMER.

In contrast to the earlier study, the best performing sample was that anodized at 50 mA/cm^2 and thus not the lowest current density. An explanation for this lies in the fact that the deep trench channel structure obtains a gradient in porosity and pore depth along the height of the channel wall, clearly seen in Figure 1.1-8b. The reactor performance is thereby characterized by a nonhomogeneous pore layer which is dependent on both the geometry of the channel structure and the anodization conditions. Figure 1.1-8a shows a pore layer thickness in the order of only 1–2 µm along the full height of the channel wall inside,

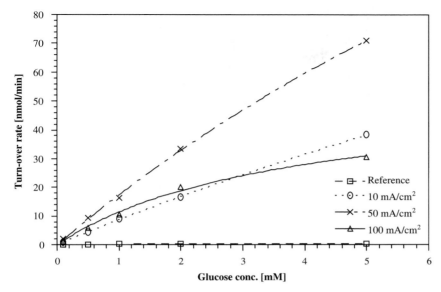

Figure 1.1-9. Glucose turnover for the three μIMERs anodized at (◯) 10, (×) 50 and (△) 100 mA/cm^2.

whereas in Figure 1.1-8c the pore layer thickness varies from approximately 50 μm at the channel bottom to 10 μm at the top of the channel wall inside. It should also be noted that the μIMERs were fabricated in p-type silicon (10–20 Ω cm), which makes direct comparisons with the initial experiments on n-type PS samples difficult.

The best PS μIMERs (50 mA/cm^2) were subsequently used in a flow injection analysis (FIA) system for glucose monitoring, injecting 0.5 μL glucose samples on the μIMER at a flow rate of 25 μL/min. An in-house-developed Clarke-type oxygen electrode [27] was connected to the outlet of the μIMER, monitoring the change in dissolved oxygen as the glucose passed through the microreactor and was converted to gluconic acid and hydrogen peroxide. Figure 1.1-10 shows the measurement system set-up. Figure 1.1-11 shows the calibration plot obtained for the glucose monitoring system. A more detailed description of this investigation is given in [28].

1.1.6.1 Influence of Silicon Dopant Type

As reported earlier, the dopant type and level strongly influence the pore morphology obtained [29] and consequently the enzyme activity achieved should vary considerably between differently prepared PS samples. In order to search for optimal porous conditions for PS as a surface for enzyme immobilization, a study involving a range of different silicon types was performed [30] (Table 1.1-2). Three p-

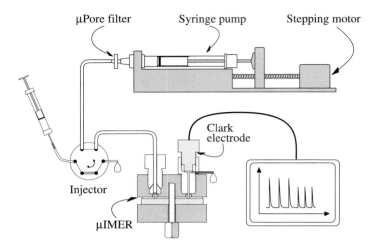

Figure 1.1-10. Glucose monitoring system FIA set-up using glucose-activated PS μIMERs and an in-house developed micro Clark electrode.

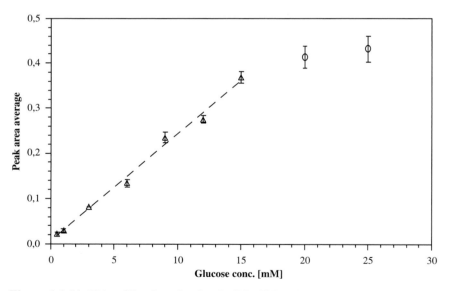

Figure 1.1-11. FIA calibration plot for the PS μIMER-based glucose monitoring microsystem.

Table 1.1-2. Silicon samples selected for the investigation of the influence of dopant type and dopant concentration on pore morphology

Sample	Specification
A	n-Epilayer on n$^+$ (6.8–9.2/0.015 Ω cm)
B	p$^+$ type (0.001–0.025 Ω cm)
C	p$^+$ type (8–12 Ω cm)
D	p$^+$ type (10–20 Ω cm)
E	Polysilicon (p substrate)

Table 1.1-3. Anodization conditions for samples A–E and the porous depth obtained for each current density applied

Sample type	Current density (mA/cm^2)	Anodization time (min)	Obtained porous depth (μm)
A1	10	10	9
A2	50	5	17
A3	100	5	24
B1	10	10	11
B2	50	5	21
B3	100	5	33
C1	10	10	~1
C2	50	5	4
C3	100	5	7
D1	10	10	11
D2	50	5	29
D3	100	5	45
E1	10	1	~1
E2	50	0.25	~1
E3	100	NA	NA

type samples, one n-type sample, and a polysilicon sample (p-type), anodized at different current densities and times, were investigated (Table 1.1-3).

The samples were anodized in a circular region of 13.5 mm diameter and during the measurement an in-house built flow cell was used for the subsequent measurements.

The previously used protocol for glucose oxidase immobilization and activity monitoring was employed. The maximum apparent turnover rate $V_{max\,app}$ for each sample surface was estimated by fitting the data according to a first-order enzyme kinetic system. Figure 1.1-12 shows the gain in catalytic turnover of the PS samples as compared with the corresponding nonporous sample. The highest increase was obtained on sample A3, which displayed a 350-fold increase in catalytic turnover. The second best performing substrate was sample D3, with a catalytic increase of 225-fold. After SEM analysis of the PS samples it was con-

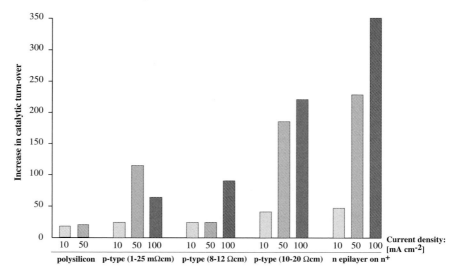

Figure 1.1-12. Increase in catalytic turnover for PS samples A–E anodized at 10, 50 and 100 mA/cm^2.

cluded that a combination of macropores at the micrometer level and mesopores in the 10–100 nm range is the common denominator for the two best performing samples A and D. The large micrometer-sized pores serve as rapid transport routes for substrate and products into the full depth of the porous layer, which is composed of a spongy, high surface area porous matrix of enzyme-activated nanometer pores that provide a high catalytic turnover. Figure 1.1-13 a and b show SEM images of samples A and D anodized at 100 mA/cm^2.

Since sample A was an epilayer substrate and therefore could not easily be adapted to the desired microchannel IMER structure, the second best performing substrate, sample D, was used for further studies of PS μIMER development. This silicon was a natural selection for the subsequent work since it also was available in ⟨110⟩ orientation, which was used previously for etching the deep channel reactor (see Section 1.1.2).

1.1.6.2   Influence of Porous Depth

As observed in the first studies on PS μIMERs (Figure 1.1-8), the depth of the PS layer varies considerably in the microreactor channel. With a thicker PS layer the diffusion-driven transport of substrate and products to the deeper catalytic regions becomes slower and hence these regions become less efficient. A too thick PS layer will not add any further catalytic performance to the μIMER and possibly only induce band broadening when using the reactor in a FIA mode.

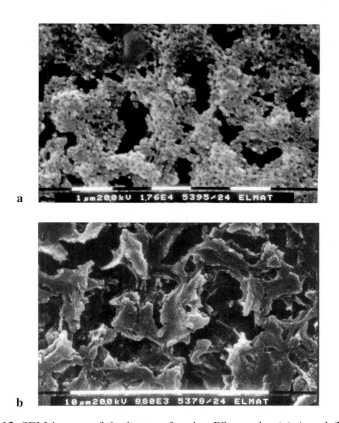

Figure 1.1-13. SEM images of the best performing PS samples (**a**) A and (**b**) D.

In order to find the optimal depth of the porous layer, a study was undertaken to investigate both planar and µIMER structures with varying pore depth and anodized under different current densities [31]. The silicon used was ⟨110⟩ silicon, p-type (10–20 Ω cm). The samples (planar and reactors) were anodized at 10 or 50 mA/cm^2, each at three different anodization times, to provide samples with a range of porous depths. At higher current densities the anodization time was shorter since the pore propagation rate in the PS etching is strongly dependent of the current density [19]. The pore depth of the planar samples was obtained by SEM measurements. Since the µIMERs displayed various pore depths along the height of the microchannels, an average value of the pore depth was derived from the scanning electron micrographs by measuring the porous depth at positions A–D in the microchannel (Figure 1.1-14).

Enzyme activities were monitored by the previously used glucose oxidase protocol [13]. Figure 1.1-15 shows the enzyme activities obtained for each of the sample types. The planar samples displayed a negligible increase in catalytic activity as the pore depth grew deeper than 20 µm and thus approximately 20 µm is the recommended depth of the PS layer when working with planar catalytic

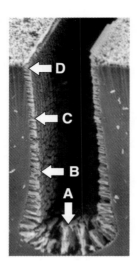

Figure 1.1-14. The average depth of the nonuniform pore layer was calculated from depth measurements taken at the locations marked A–D.

surfaces. The μIMER displayed a much shallower pore depth at the point where the increase in activity leveled off. This was explained by the fact that the actual pore depth was given by an average of the depth measured at the channel bottom and at three points along the height of the channel wall. As is observed in the scanning electron micrographs (Figure 1.1-8b and c), the porous layer is much thinner on the channel walls than at the bottom. Owing to the electric field-enhanced attraction of holes to the channel bottom, very few holes pass this region and diffuse further up the channel wall to generate pores in this region. As the anodization process proceeds, the pore depth continues to grow at the channel bottom whereas the growth rate of the porous layer on the channel wall is reduced as the porous region at the channel bottom broadens and eventually consumes all bulk silicon that can provide hole transport to the channel walls (Figure 1.1-16). The major part of the μIMER geometric surface area is obtained by the deep channel walls and as the pore depth grows deeper and beyond the efficient depth at the channel bottom, the PS depth at the channel walls in principle remains unchanged. Consequently, the data for the reactors in Figure 1.1-15 indicate a too shallow optimal porous depth. To utilize better the geometric area available for porosification, it would therefore be desirable to obtain more homogeneous pore depths along the height of the channel. This has also been the focus of more recent studies (see Section 1.1-11).

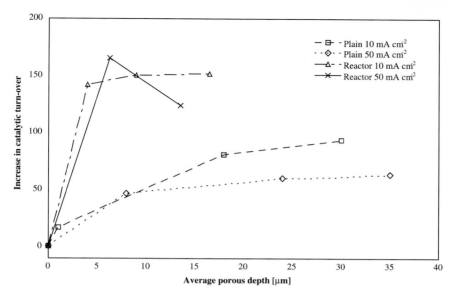

Figure 1.1-15. Average pore depth vs increase in enzyme activity for planar samples and µIMERs anodized at 10 and 50 mA/cm² and with different porous depths.

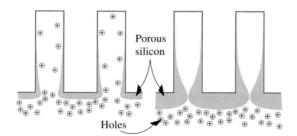

Figure 1.1-16. Schematic diagram of the broadening of the porous layer at the channel bottom as the anodization proceeds, which stops the further migration of holes from the bulk silicon into the channel wall region. This arrests the further growth of the PS layer on the channel walls.

1.1.7   Sucrose Analysis

The best versions of the PS µIMERs as found in the investigations performed earlier [28] (see Section 1.1.5) were later involved in the development of new applications to monitor the sucrose content in soft drinks by Lendl et al. [32]. The system was based on a PS µIMER with β-fructosidase immobilized in the porous matrix according to previously used protocols. The µIMER selectively hydrolyzed sucrose to fructose and glucose and the reactor effluent was coupled on-

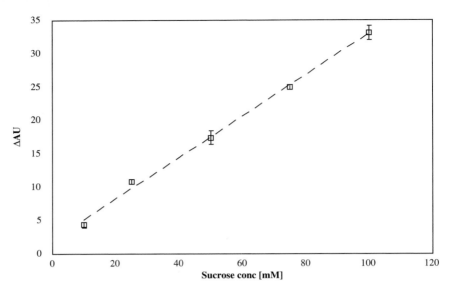

Figure 1.1-17. Calibration plot for sucrose using the µIMER/FTIR-based chemical micro-analysis system.

line to a fiber-optic microflow cell of 10 nL in which Fourier transform infrared (FTIR) spectroscopy was performed on-line. By recording the FTIR spectra of both the hydrolyzed and unhydrolyzed sucrose samples, via a flow line bypassing the µIMER, a differential spectrum was obtained from which the changes in absorbance at sucrose characteristic wavenumbers were calculated. The system was calibrated on sucrose standard solutions of concentration 10–100 mM (Figure 1.1-17). It was also shown that the absorbance measurements at the selected wavenumbers were not affected by the presence of glucose and thus did not interfere with the sucrose determination. Subsequent studies also comprised the analysis of the sucrose content in crude soft drink samples. The microsystem analysis displayed excellent performance when compared with data derived from standardized sucrose measurements (Figure 1.1-18).

1.1.8   Glutamate Monitoring

In neuroscience research, much effort is aimed at acquiring a better understanding of neurological diseases such as Parkinson and Alzheimer disease and schizophrenia. Glutamate plays an important role in neural signal transduction and is thus a compound of quantitative and temporal interest in vivo. In order to perform real-time measurements of glutamate in brain, noninvasive or semi-noninva-

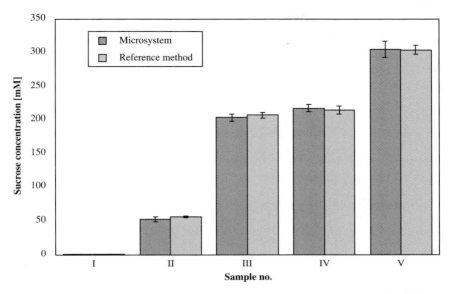

Figure 1.1-18. Comparison of sucrose measurements performed on crude soft drink samples using either the developed μIMER/FTIR microsystem or standardized analysis protocols.

sive microbiosensors have to be developed. Commonly glutamate is recorded by means of electrochemical measurements in which glutamate oxidase catalyzes the conversion of glutamate with the production of H_2O_2. The hydrogen peroxide is then electrochemically recorded either via direct oxidation at the electrode surface or through an enzymatic conversion via peroxidase that is 'wired' to the electrode surface via an electropolymer [33, 34], ensuring that the electron transport in the redox process is shuttled via the electrode. The wired peroxidase electrode allows for a lower electrode potential, thereby reducing influences of electrochemical interference. At this potential ascorbate is still a major interferent, present at levels more than 50 times higher than the targeted analyte glutamate.

In an effort to address this, we developed a microsystem for glutamate monitoring which is based on microdialysis sampling linked on-line to a wired glutamate oxidase/peroxidase biosensor. A detailed system description can be found in [35]. In order to eliminate the influence of ascorbate, an interferent preoxidation step was implemented using an ascorbate oxidase (AscOx) activated PS μIMER. Figure 1.1-19 shows the monitoring system where the sample during the system development was provided via bulk solution injections and later from brain microdialysis. The volumes injected were 0.2 μL and, as the sample proceeded through the AscOx μIMER, the interferent was oxidized prior to reaching the wired glutamate/peroxidase biosensor. Figure 1.1-20 shows the calibration plot for the glutamate biosensor (solid line) and the interferent effect (dashed line) of 100 μM ascorbate, which is a relevant interferent level for in vivo samples. It should be noted that typical glutamate levels are in the range of 10 μM. In order to reduce the decrease in glutamate sensitivity due to dispersion as the sample passed the μIMER, two reactor dimensions (6

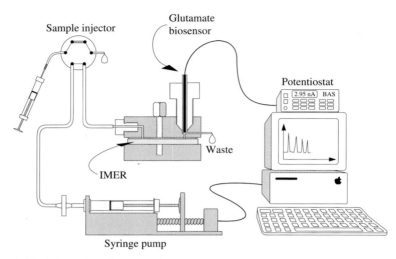

Figure 1.1-19. Schematic diagram of the glutamate monitoring system with the ascorbate preoxidation PS µIMER.

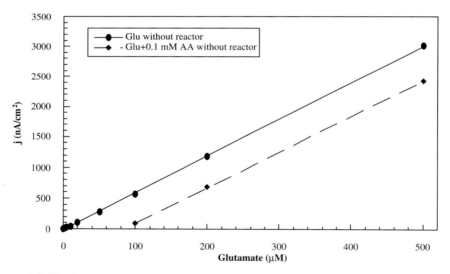

Figure 1.1-20. Glutamate sensor calibration with (dashed line) and without (solid line) 100 µM ascorbate added to the glutamate standard.

and 1 µL internal volume) were investigated. At flow rates below 25 µL/min, the 6 µL reactor eliminated all ascorbate in a 1 mM sample, whereas the smaller µIMER displayed an elimination level of 0.2 mM at a flow rate of ≤10 µL/min.

In the subsequent testing of the glutamate biosensor system, the 6 µL AscOx preoxidation µIMER was used. Figure 1.1-21 shows the glutamate calibration

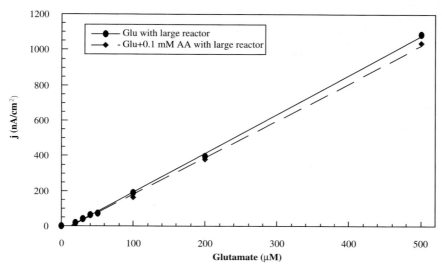

Figure 1.1-21. Calibration plot for the glutamate monitoring system, using an AscOx-activated PS µIMER for interference elimination, without (dashed) and with (solid) the addition of 100 µM ascorbate.

plot for pure glutamate samples (solid line) and with 100 µM ascorbate added (dashed line). No influence of the interference could be seen. The minor difference in slope of the calibration plot was attributed to the fact that two different biosensors were used for each of the measurements.

1.1.9   µIMER Applications in Protein Analysis

The announcement of the completion of the human genome in the summer of 2000 by Celera and the Human Genome Project [36] automatically served as the starting point for the next major effort in the global life science research community, i.e., the mapping of all proteins that are expressed in a biological system at any time during, eg, progress of a disease. In this process, there is a great need for new and highly efficient protein analysis technologies. It is anticipated that no single technical solution as in the DNA case will provide all the analytical performance needed since proteins are much more complex than DNA. However, the use of matrix-assisted laser desorption/ionization time-of-flight mass spectrometry (MALDI-TOFMS) in life science research has, since its introduction [37], evolved as a fundamental and generic technique to read protein structures [38]. In order to address the variability of proteins, a wide set of analytical techniques are being developed to separate protein mixtures in time and space, deriving fractions of single or a few proteins that subsequently can be analyzed by mass spectrometry (MS). These fractions are precious and commonly of low volume (1–10 µL), which in turn has driven the rapid development of new microanalytical techniques for the efficient sample processing prior to the MS analysis.

In general the protein is subjected to a digestion step in which a protease cleaves the protein into peptide fragments, providing a peptide mixture of predetermined sizes for a specific protein. The protein digestion typically runs for a few hours or overnight. The peptide mixture is then determined by MS and the peptide map mass fingerprint obtained is used for database searching in order to find the matching protein.

In an effort to improve the conditions for performing protein digestion in proteomics research, our group has developed PS µIMERs that are activated with proteases for rapid on-line protein digestion [39]. The reactor initially used in these applications was of the same geometry as investigated and optimized in Section 1.1.5, glucose monitoring [28]. Figure 1.1-22 shows the miniaturized on-line protein digestion/analysis platform. Samples of 1 µL were injected into the carrier flow and as the sample passed the trypsin-activated PS µIMER, the protein was subjected to digestion. The digested sample then proceeded to a flowthrough microdispenser [40] from which the sample was microdispensed on to microfabricated nano-vial MALDI target arrays [41]. The MALDI targets were subsequently analyzed by MALDI-TOFMS and peptide fingerprint database mapping. It was found that a surprisingly short residence time for the protein sample in the PS µIMER was needed. At a flow rate of 6 µL/min, ie, 60 s residence time in the µIMER, a suffi-

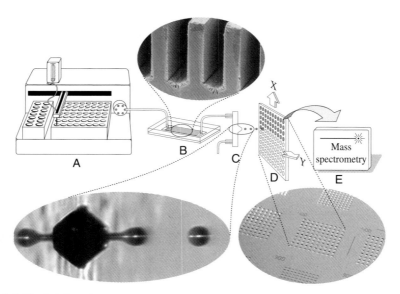

Figure 1.1-22. Miniaturized protein digestion and MALDI sample handling system. (**A**) Sample handling robot; (**B**) protease-activated µIMER; (**C**) microdispenser; (**D**) nano-vial target plate; (**E**) mass spectrometric analysis.

cient digestion for unambiguous database identification was found for the model proteins investigated, myoglobin and lysozyme. Figure 1.1-23 shows mass spectra obtained for lysozyme (1 pmol/µL) and myoglobin (3 pmol/µL), digested on a (A) trypsin- and a (B) chymotrypsin-activated PS µIMER, respectively. An attractive feature with the miniaturized approach is the possibility of splitting the original

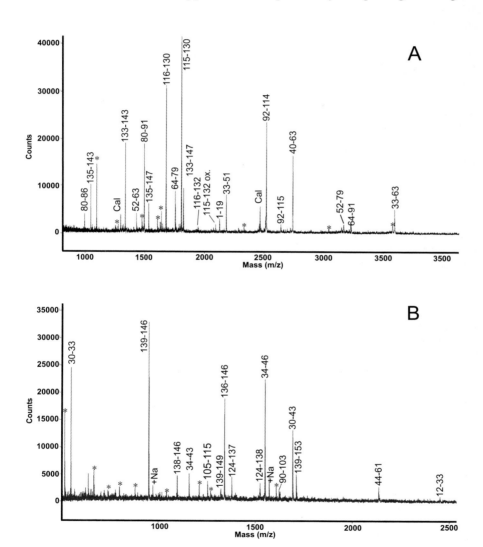

Figure 1.1-23. Mass spectra obtained from (**A**) a 1 pmol/µL sample of lysozyme digested on a trypsin PS µIMER giving a sequence coverage of 91% and (**B**) a 3 pmol/µL myoglobin sample digested on a chymotrypsin PS µIMER yielding a sequence coverage of 68%. Cal, internal calibrants; ox, methionine residue oxidized; *, unidentified peaks; +Na, sodium adducts.

sample into several 1 μL sub-fractions, each fraction passing through a μIMER with a different protease. Thereby different cleavage patterns are obtained for the same sample, which increases the amount of protein-specific data and thus the probability of a successful database search is improved.

The protease-activated PS μIMER provides a highly catalytic microenvironment for proteins which yields short digestion times. The described system reduced the standard digestion protocols that run for 6–24 h to only 60 s. The system was shown to process 100 samples in 300 min. The sample volumes required could be reduced by a factor of 10–100 compared with conventional techniques, thereby enabling sample subfractionation and multiple enzyme digestion. The described microsystem has demonstrated the benefits of miniaturization, and especially the gains that are obtained by including porous silicon microstructures as a carrier matrix for immobilized proteases are evident.

1.1.10 Reactor Stability

When immobilizing enzymes, the stability of the derived catalytic component is of great importance. In the course of studying the influence of differently doped silicon and the subsequent outcome of varying pore morphologies (see Section 1.1.6.1), the best performing sample (A3, Table 1.1-3) was stored at 8 °C for 5 months, after which the enzyme activity remaining was monitored to obtain a value of the storage stability of the immobilized enzyme. The sample was thereafter continuously exposed to glucose for 4 days, during which the remaining enzyme activity was recorded daily. Figure 1.1-24 shows the storage and operational stability of the immobilized enzyme. The storage stability measurement indicated a 2–

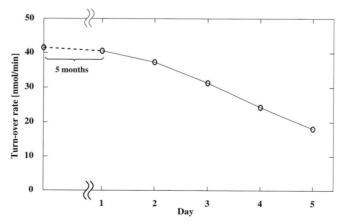

Figure 1.1-24. Storage (5 months) and operational (4 days) stability of the PS μIMERs.

3% decrease in activity during the 5 month period. The operational stability measurement showed an activity decrease of 8% on the first day, whereafter the decrease was approximately 15% per day. Keeping in mind that the operational stability measurement was performed in a continuous mode and that the application for the µIMER is as a component in micro-FIA systems, the operational stability should be expected to be considerably improved since the enzyme is only exposed to substrate, as the injected pulse passes through the µIMER. This is also supported by the applications performed on protein digestion described above, where reactors were used for several weeks, digesting approximately 400 samples without displaying any reduced protein digestion performance.

1.1.11 Recent Developments

In order to optimize further the catalytic performance of the PS µIMER, studies have been focused on obtaining a more homogeneous porous layer along the full height of the channels in the µIMER. As seen in Figure 1.1-14, the porous depth at the channel bottom varies considerably and also the pore structure displays a dramatic change in morphology. This is clearly seen in the side view SEM of the channel wall from a µIMER anodized at 50 mA/cm^2 (Figure 1.1-25).

By obtaining a more homogeneous PS layer with correct morphology on the channel wall, a further improvement in catalytic performance was expected. This can be obtained by redesigning the microreactor such that holes can diffuse through bulk silicon to the higher regions of the channel walls. By increasing the wall width/channel width from the 50/50 µm ratio used earlier, more bulk silicon is obtained in

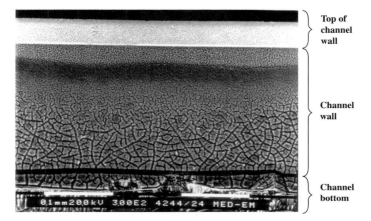

Figure 1.1-25. Side view of a channel wall from a µIMER anodized at 50 mA/cm^2, showing the pore morphology gradient resulting from the nonhomogeneous pore formation conditions along the height of the channel wall during the anodization.

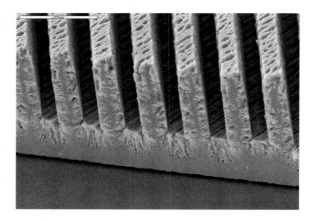

Figure 1.1-26. Cross-section of a PS µIMER having a homogeneous pore depth along the full height of the channel walls, thereby improving the catalytic performance of the µIMER.

the channel wall. Recent data have verified that this is a plausible strategy and an increase in reactor turnover was obtained for reactors with 75/25 µm wall/channel ratio as compared with the conventional 50/50 µm ratio [42]. These reactors have also recently been applied to the proteomics application described above, demonstrating an even higher protein digestion rate than the 60 s previously reported [43]. Figure 1.1-26 shows a cross-section of the PS µIMER with a more homogeneous pore depth along the full height of the reactor channel walls.

1.1.12 Conclusions

PS has been shown to be an excellent surface area-enlarging biocatalytic matrix that can easily be integrated with chemical microsystems. Surface modification protocols for PS to couple a wide variety of bioactive compounds can be derived from the wealth of knowledge on surface modifications of silica matrices frequently used in biochemistry. This opens up an immense area of applications for miniaturized bioactive analysis systems. The examples given in this review indicate only a few of the application areas for enzyme-activated PS. Ongoing work is investigating the combination of cells immobilized in PS µIMERs with immobilized enzymes to provide integrated multi-step processing of complex analytes [44]. The role of miniaturization is a key feature to reduce the sensitivity decrease due to sample dispersion. Also, the fact that microfluidic structures such as the deep-channel µIMER can be combined with a surface-enlarging PS matrix to bring high surface area components into microsystems makes the integration of efficient bioactive components at the chip level a highly attractive analytical concept.

1.1.13 References

[1] Manz, A., Grabner, N., Widmer, H. M., *Sens. Actuators B* **1** (1990) 244–248.
[2] Van den Berg, A., Bergveld, P. (eds.), *Micro Total Analysis Systems, Proceedings of the µTAS 94 Workshop;* Dordrecht: Kluwer, 1994.
[3] Widmer, H. M., Verpoorte, E., Barnard, S. (eds.), *Proceedings of the 2nd International Symposium on Micro Total Analysis Systems, µTAS '96; Analytical Methods and Instrumentation,* Special Issue, 1996.
[4] Harrison, J. D., van den Berg, A. (eds.), *Micro Total Analysis Systems 98, Proceedings of the 3rd International Symposium on Micro Total Analysis Systems;* Dordrecht: Kluwer, 1998.
[5] Van den Berg, A., Olthuis, W., Bergveld, P. (eds.), *Micro Total Analysis Systems 2000, Proceedings of the µTAS 2000 Symposium;* Dordrecht: Kluwer, 2000.
[6] Weetall, H. H., *Methods Enzymol.* **44** (1976) 134–148.
[7] Suda, M., Sakuhara, T., Murakami, Y., Karube, I., *Appl. Biochem. Biotechnol.* **41** (1993) 11–15.
[8] Murakami, Y., Toshifumi, T., Yokoyama, K., Tamiya, E., Karube, I., Suda, M., *Anal. Chem.* **65** (1993) 127–130.
[9] Xie, B., Danielsson, B., Norberg, P., Winquist, F., Lundström, I., *Sens. Actuators B* **6** (1992) 127–130.
[10] Danielsson, B., *Ann. N. Y. Acad. Sci.* **501** (1987) 543–544.
[11] Strike, D. J., Thiébaud, P., van der Sluis, A. C., Koudelka-Hep, M., de Rooij, N. F., *Microsyst. Technol.* **1** (1994) 48–50.
[12] Laurell, T., Rosengren, L., *Sens. Actuators B* **19** (1994) 614–617.
[13] Laurell, T., Rosengren, L., Drott, J., *Biosens. Bioelectron.* **10** (1995) 289–299.
[14] Laurell, T., *PhD Thesis*, Department of Electrical Measurements, Lund Institute of Technology, 1995.
[15] Uhlir, A., *Bell Syst. Tech. J.* **35** (1956) 333–347.
[16] Watanabe, Y., Arita, Y., Yokoyama, T., Igarashi, Y., *J. Electrochem. Soc.: Solid-State Sci. Technol.* **122** (1975) 1351–1355.
[17] Bomchil, G., Barla, K., Herino, R., *J. Electrochem. Soc.* **13** (1986) C96–C96.
[18] Canham, L. T., *Appl. Phys. Lett.* **57** (1990) 1046–1050.
[19] Smith, R. L., Collins, D. C., *IEEE Trans. Biomed. Eng.* **33** (1986) 83–90.
[20] Laurell, T., Drott, J., Rosengren, L., Lindström, K., *Sens. Actuators B* **31** (1996) 161–166.
[21] Thust, M., Schoning, M. J., Frohnhoff, S., ArensFischer, R., Kordos, P., Luth, H., *Meas. Sci. Technol.* **7** (1996) 26–29.
[22] Schoning, M. J., Ronkel, F., Crott, M., Thust, M., Schultze, J. W., Kordos, P., Luth, H., *Electrochim. Acta* **42** (1997) 3185–3193.
[23] Smith, L., Collins, D. C., *J. Appl. Phys.* **71** (1992) R1–R22.
[24] Nassiopoulos, A. G., in: *Properties of Porous Silicon*, Canham, L. (ed.); pp. 77–80.
[25] Eijkel, C. J. M., Branebjerg, J., Elwnspoek, M., Vandepol, F. C. M., *IEEE Electron Dev. Lett.* **11** (1990) 588–589.
[26] Von Gallati, H., *J. Clin. Chem. Clin. Biochem.* **15** (1977) 699–703.
[27] Laurell, T., *Sens. Actuators B* **13/14** (1993) 323–326.
[28] Drott, J., Lindström, K., Rosengren, L., Laurell, T., *J. Micromech. Microeng.* **7** (1997) 14–23.

[29] Herino, R., Bomchil, G., Barla, K., Bertrand, C., *J. Electrochem. Soc.* **134** (1987) 1994–2000.
[30] Drott, J., Rosengren, L., Lindström, K., Laurell, T., *Thin Solid Films* **330** (1998) 161–166.
[31] Drott, J., Rosengren, L., Lindström, K., Laurell, T., *Mikrochim. Acta* **131** (1999) 115–120.
[32] Lendl, B., Schindler, R., Frank, J., Kellner, R., *Anal. Chem.* **69** (1997) 2977–2991.
[33] Vreeke, M., Maidan, R., Heller, A., *Anal. Chem.* **64** (1992) 3084–3090.
[34] Belay, A., Collins, A., Ruzgas, T., Kissinger, P.T., Gorton, L., Csöregi, E., *J. Pharm. Biomed. Anal.* **19** (1999) 93–105.
[35] Collins, A., Mikeladze, E., Bengtsson, M., Laurell, T., Csöregi, E., *Electroanalysis* **13** (2001) 425–431.
[36] Pennisi, E., *Science* **288** (2000) 2304–2307.
[37] Karas, M., Bachman, D., Bahr, U., Hillenkamp, F., *Int. J. Mass Spectrom. Ion Processes* **78** (1987) 53–68.
[38] Chalmers, M.J., Gaskell, S.J., *Curr. Opin. Biotechnol.* **11** (2000) 384–390.
[39] Ekström, S., Önnerfjord, P., Nilsson, J., Bengtsson, M., Laurell, T., Marko-Varga, G., *Anal. Chem.* **72** (2000) 286–293.
[40] Laurell, T., Wallman, L., Nilsson, J., *J. Micromech. Microeng.* **9** (1999) 369–376.
[41] Ekström, S., Ericsson, D., Önnerfjord, P., Bengtsson, M., Nilsson, J., Marko-Varga, G., Laurell, T., *Anal. Chem.* **73** (2001) 214–219.
[42] Bengtsson, M., Drott, J., Laurell, T., *Phys. Status Solidi A* **182** (2000) 533–539.
[43] Bengtsson, M., Ekström, S., Marko-Varga, G., Laurell, T., *Talanta* (2001) in press.
[44] Davidsson, R., Bengtsson, M., Johansson, B., Passoth, V., Laurell, T., Emneus, J., in: *Micro Total Analysis Systems 2001, Proceedings of the µTAS 2001 Symposium.*

List of Symbols and Abbreviations

Symbol	Designation
V_{max}	maximum substrate turnover rate
$V_{max\ app}$	maximum apparent turnover rate

Abbreviation	Explanation
AscOx	ascorbate oxidase
EIS	electrolyte insulator semiconductor
FIA	flow injection analysis
FTIR	Fourier transform infrared
MALDI-TOFMS	matrix-assisted laser desorption/ionization time-of-flight mass spectrometry
µIMER	micro immobilized enzyme reactor
µTAS	micro total analysis
PS	porous silicon
PVDF	poly(vinyl difluoride)
SEM	scanning electron microscopy

1.2 Construction of a Living Coupled Oscillator System of Plasmodial Slime Mold by a Microfabricated Structure

A. TAKAMATSU, PRESTO, Japan Science and Technology Corporation (JST), Saitama, Japan
T. FUJII, Institute of Industrial Science, The University of Tokyo, Tokyo, Japan

Abstract

The cell patterning technique is important for observing living cells in a systematic way, especially from the viewpoint of nonlinear science, which enables us to treat a complicated system such as a living system by a form of mathematics. We developed a cell patterning method and constructed a living coupled oscillator system with the plasmodial slime mold, *Physarum polycephalum*, using a microfabricated structure.

The plasmodium is a large ameboid unicellular organism consisting of an almost homogeneous structure without a highly differentiated system like a nervous system, but it shows a response to change in environment. The response behavior is considered to be caused by the interactions among the cell bodies where internal nonlinear oscillatory phenomena can be observed. Therefore, the plasmodium can be regarded as a coupled oscillator system. In order to understand the mechanism of the behavior of the plasmodium, it is essential to observe the behavior under conditions where the interactions among the cell bodies are systematically controlled.

We patterned the shape of the plasmodium by a microstructure fabricated with a photoresist resin (SU-8) by a photolithographic method. With this system, we constructed two- and multiple oscillator systems of the plasmodium and observed oscillation phenomena by controlling the interaction parameters among the oscillators. We found that the oscillation patterns varied dynamically depending on the parameters in the same manner of the prediction by a mathematical model of a coupled oscillator system.

This approach with side-by-side analyses is important for understanding complicated biological systems. In addition, for such analyses, the microfabrication technique is one of the key technologies.

Keywords: Coupled oscillator system; nonlinear dynamics; *Physarum polycephalum*; microfabrication; SU-8; indium tin oxide

Contents

1.2.1   A Scientific Application of the Microfabrication Technique – Construction of a Living Coupled Oscillator System

The cell patterning method is an important technique for observing living cells in a systematic way, especially from the viewpoint of nonlinear science, which enables us to treat a complicated system such as a living system by a form of mathematics. Further, the microfabrication technique gives us methods to control the shapes of small cells (10–100 µm). In this study, we developed a cell patterning method with a microfabricated structure and constructed a living coupled oscillator system of the plasmodial slime mold *Physarum polycephalum*.

Oscillatory phenomena are generally observed in every hierarchy in biological systems from a cell or a tissue to a society of living organisms, for example, in activities of electrical potential of neuronal cells and cardiac cells, circadian rhythms, and the collective flashing of fireflies [1, 2]. Systems generating oscillation have two characteristics: robustness, ie, even if the oscillation element is perturbed by an external stimulus, the system instantaneously recovers; and cooperativity, ie, the collective of the element can synchronize with the same rhythm through the interactions among the elements even if the original rhythms of the individual elements differ from one another [3, 4]. The characteristics of these phenomena can be commonly observed not only in biological systems but also in a class of system, so-called coupled nonlinear oscillator systems. The coupled oscillator system is a collective of units (nonlinear oscillators) that generate rhythm, and the oscillators interact (couple) with each other in some way, eg, synaptic connections, chemical diffusion. The generalized theory of the coupled oscillator system is based on simple and fundamental mathematical models, and it gives us perspectives to understand complicated biological systems. The theory is, however, too simple to be compared directly with real biological systems. Therefore, we propose a living coupled oscillator system, namely a model system for mathematical analysis built with the living cell, by reconstructing the biological system as a collective of elements using a cell patterning method.

Other researchers have adopted the same approach. For example, they have tried to record cell activities such as neurons under the condition that the geometry or configuration of the cells is controlled [5–7]. Experimental treatment of the cells such as neurons is difficult, however, because of the delicate characteristics of the cells, and some technical problems remain for research on collective dynamics. We therefore adopted a cell whose treatment is relatively easy, the plasmodial slime mold, to realize the construction of a living coupled oscillator system.

1.2.2 The Plasmodial Slime Mold

The plasmodium of the slime mold (Figure 1.2-1 A) *Physarum polycephalum* is an amoeboid multinucleated unicellular organism and is an aggregate of protoplasm without any highly differentiated structure like a nervous system. The plasmodium can be divided into multiple parts that are kept alive. On the other hand, the multiple parts of the plasmodium can fuse into one cell. That is, any part of the plasmodium is almost identical and the plasmodium itself is merely a collective of those partial bodies.

Despite its simple structure, the plasmodium demonstrates sophisticated biological functions. A large cluster of the plasmodium can behave as a single entity through the interactions among the partial bodies. When external stimuli are locally applied, the plasmodium exhibits gathering/escaping behavior to/from attractants/repellents by transmitting the locally acquired information throughout the whole cluster [8]. Thus, it can be said that the plasmodium has a highly distributed information processing mechanism.

These biological functions can be understood through the intrinsic nonlinear oscillatory phenomena in the plasmodium, eg, oscillations in concentration of intracellular chemicals such as ATP [9] and Ca^{2+} [10], and contraction/relaxation rhythm [11] driving thickness oscillation in the plasmodium [12]. These oscillatory phenomena are considered to be generated by complicated mechano-chemical reactions among the chemicals, organelles, and proteins in the cell [13]. Since a unit generating oscillatory phenomena can be defined as an oscillator, the partial bodies of the plasmodium can be treated as the plasmodial oscillators.

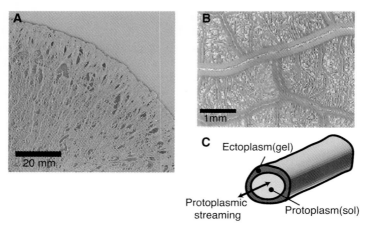

Figure 1.2-1. (A) The plasmodium of the slime mold *Physarum polycepharum*. **(B)** Tube structure in the plasmodium. **(C)** Schematic diagram of the tube structure.

They are interconnected by a tube structure (Figure 1.2-1 B and C), in which the protoplasm streams periodically owing to the pressure difference generated by the contraction/relaxation rhythm [14]. The plasmodial oscillators are, therefore, considered to be coupled through the protoplasmic streaming. Hence, by systematically controlling the dimension of the tube structure that would correspond to the coupling parameters, it would be possible to observe a variety of oscillatory phenomena. The results would help us to understand the behavior of the plasmodium from the viewpoint of nonlinear mathematical analysis.

1.2.3   Two-oscillator System

As the first step, we constructed a two-oscillator system with the plasmodial slime mold using a microfabrication technique. In this section, we describe a method for constructing the system, then basic characteristics of the coupled two-oscillator system of the plasmodial slime mold.

1.2.3.1   Cell Patterning

In order to construct a living coupled oscillator system, the plasmodial slime mold is patterned by a microfabricated structure. This is made of ultrathick photoresist resin (NANO^{TM} SU-8 50, Microlithography Chemical) and is fabricated through a conventional photolithography method [15]. The structure is a thick sheet, about 100 μm in thickness, and has a dumbbell-shaped opening as

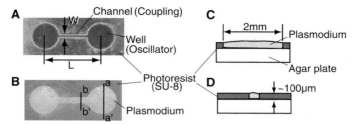

Figure 1.2-2. Two-oscillator system. (**A**) Microfabricated structure. Diameter of the wells, 2 mm. Channel width, denoted by *W*, 50–800 μm. Channel length, denoted by *L*, 3–15 mm. (**B**) The plasmodium patterned by the microfabricated structure. (**C**) Well (oscillator) part; a–a' section in B. (**D**) Channel (coupling) part; b–b' section in B. (Adapted from [15, 16]).

shown in Figure 1.2-2 A. The structure is placed on a 1.5% agar plate, on which the plasmodium is usually cultured, to pattern the wet (agar) and dry (microfabricated structure) surfaces (Figure 1.2-2 B–D). The plasmodium spreads only on the wet surface since it prefers a wet to a dry surface so it can be patterned according to the shape of the opening in the microfabricated structure as shown in Figure 1.2-2 B.

1.2.3.2 Construction of Two-oscillator System and Controlled Parameters

The microfabricated structure for two-oscillator system consists of wells and a channel, which correspond to oscillators and a coupling, respectively (Figure 1.2-2 A). Small pieces of plasmodia are put in the wells. The plasmodium in the microfabricated structure is cultured in a thermostat and humidistat chamber (PR-2K, ESPEC) at $25\pm0.3\,°C$ and RH 75–90%. After 5–10 h, the plasmodia in both wells grow along with the channel structure and contact each other around the center of the channel. Then tube structure is formed along the channel [15]. Finally, the plasmodia in the wells are physically connected by the tube structure and the protoplasmic streaming can be observed inside. We observed thickness oscillation in the well, whose diameter was determined as 2 mm. Inside the well, oscillation pattern is coherent so that we can consider the plasmodium in the oscillator part as a single oscillator.

The coupling parameters between the plasmodial oscillators can be systematically controlled by changing channel width (*W*) and channel length (*L*) of the microfabricated structure. In order to verify the fact that these parameters are properly controlled, we measured the tube diameter and the velocity of the protoplasmic streaming in the tube structure.

In the narrow channel, a single tube structure is formed and the tube diameter can be systematically controlled by the channel width as shown in Figure 1.2-3 A [15].

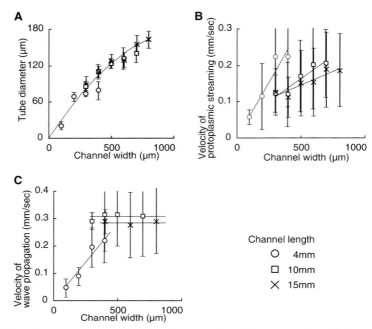

Figure 1.2-3. Parameters controlled by the channel width in the microfabricated structure of the two-oscillator system. (**A**) Tube diameter. (**B**) Time-averaged absolute value of the velocity of protoplasmic streaming in the tube structure. (**C**) Time-averaged velocity of wave propagation in the tube structure.

When single tube structures are formed, the tube diameters are dependent only on W, and not on L. However, in wider channels exceeding a certain critical width, eg, 400 µm when $L=4$ mm, a dendritic tube structure is formed and the tube diameter is no longer systematically controlled (data not shown in Figure 1.2-3 A) [15, 16]. Therefore, in the subsequent analysis, we take into account only the samples where the single tube structures are formed.

Figure 1.2-3 B shows the time-averaged absolute velocity of protoplasmic streaming observed in the middle of the tube structure. The protoplasm streams back and forth, ie, the velocity itself oscillates from a negative to a positive value. Thus, we averaged the absolute value of the velocity. The time-averaged absolute velocity is dependent on both W and L. From this result, the rich behavior in the plasmodial oscillator system is expected to be observed by controlling W and L [16].

In addition, the wave propagation velocity can also be controlled by the channel configuration as shown in Figure 1.2-3 C, whose effect will be discussed later.

1.2.3.3 Experimental Setup

Figure 1.2-4 A shows the experimental setup for the measurement of thickness oscillation in the plasmodial oscillator system. The plasmodia grown in the microfabricated structure are set under a charge-coupled device (CCD) camera (C2400; Hamamatsu) with lenses. The image transmitted through the plasmodium (eg, Figure 1.2-5 A) is captured by the CCD. The light source is filtered with a bandpass filter (580±10 nm) to avoid the photo and thermal response of the plasmodium [17]. The image data converted by the flame grabber (LG-3; Scion) are sequentially stored in a PC (Power Macintosh G3; Apple) every 4 s, which is short enough as the sampling time against one period of oscillation in the plasmodium (1–2 min). The measurements of thickness oscillation were performed in a thermostat and humidistat chamber (PR-2K, ESPEC) at 25±0.3 °C and RH 85±2.5%.

We confirmed that the transmitted light intensity is inversely proportional to the thickness within the range of thickness oscillation as shown in Figure 1.2-4 B.

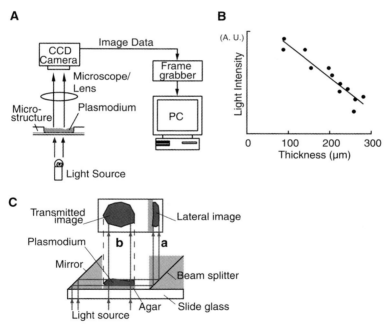

Figure 1.2-4. (**A**) Experimental setup for measurement of thickness oscillation. (**B**) Relation between thickness of the plasmodium and transmitted light intensity through the plasmodium. (**C**) Experimental setup for the simultaneous measurement of the thickness with the transmitted light intensity. The lateral image of the plasmodium is obtained via the light path (a) through the two prisms, mirror, and beamsplitter. The thickness is measured from the lateral image. The transmitted image is obtained directly through the plasmodium via the light path (b).

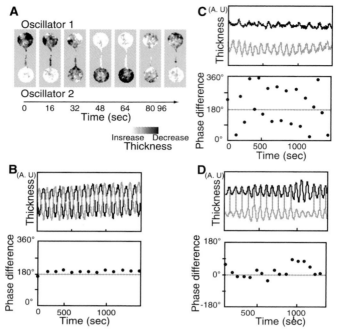

Figure 1.2-5. Oscillation patterns in a two-oscillator system. (**A**) Transmitted light images of thickness oscillation. $W=400$ µm, $L=4$ mm. (**B**)–(**D**) The upper graphs are time courses of thickness oscillation, which were obtained by averaging the light intensity in the area of each oscillator. The dark and light lines correspond to oscillators 1 and 2, respectively. The lower graphs show the phase differences between oscillators 1 and 2, which were calculated from the peak position in the waveforms. (**B**) Anti-phase oscillation. $W=400$ µm, $L=4$ mm. (**C**) Asynchronous. $W=100$ µm, $L=4$ mm. (**D**) In-phase oscillation. $W=500$ µm, $L=10$ mm. (Adapted from [16, 18])

The relationship was obtained by simultaneously measuring the transmitted image of the plasmodium with the lateral image through the two prisms (the mirror and the beam splitter) under the microscope as shown in Figure 1.2-4 C.

1.2.3.4   Oscillation Patterns Depending on the Channel Configuration

From the measurement of the thickness oscillation in the two-oscillator system of the plasmodium, various kinds of oscillatory phenomena, such as anti-phase and in-phase oscillation were found depending on the channel configuration, W and L.

When the channel width was large ($W=400$ µm) and the channel length was small ($L=4$ mm), the two oscillators synchronized in anti-phase (Figure 1.2-5 A and B). The phase difference between two oscillators is $180°$ and seems stable, as shown in Figure 1.2-5 B. When W was somewhat smaller (200–300 µm), the syn-

chronization in the two oscillators became unstable, and the system showed so-
called quasi-periodic oscillation. When W was much smaller (100 μm), the two oscillators never synchronized and oscillated at their own frequencies (Figure 1.2-5 C)
[16, 18]. In fact, under this condition, protoplasmic streaming can be observed,
which means that the two oscillators physically connected through the tube structure, even with this smaller channel width.

This phenomenon can be understood by a theory of entrainment in a coupled
nonlinear oscillator system [1, 3, 4]. The simplest mathematical model, ie, the
phase coupling model [3], shows that oscillators at different intrinsic frequencies
can be entrained to oscillate with a common frequency when the coupling
strength is larger than a critical value defined by distributions of the intrinsic frequencies. The oscillators, however, oscillate at each intrinsic frequency when the
coupling strength is lower than the critical value.

In the plasmodial oscillator system, the coupling strength would be related to
the amount of protoplasm transported by the protoplasmic streaming through the
tube structure. Since the velocity of the protoplasmic streaming was controlled
by the channel width W, as mentioned in Section 1.2.3.2, W would correspond to
the coupling strength (strictly, the coupling strength depends mainly on W but
also on L [18]).

The anti-phase oscillation in the plasmodial system can be understood as follows. The changes in the amount of protoplasm caused by the thickness oscillation is of the same order as those transported by the protoplasmic streaming
through the tube structure according to our calculations [18]. In other words, the
protoplasmic streaming would directly affect the change in the thickness of the
oscillators, which would cause a contraction in one oscillator and an expansion
in the other with a short distance connection. Therefore, the coupling between
the two oscillators can be reciprocal.

In contrast, we observed more interesting phenomena depending on the channel length L. When both the channel length and the channel width are large
($L=10$ mm, $W=500$ μm), in-phase oscillation is found, as shown in Figure 1.2-
5 D. The transition from anti-phase to in-phase oscillation cannot be explained
only by the phase coupling model mentioned above at any coupling strength. In
fact, it has been proved that the phenomenon can be explained if another parameter, the delay time, is introduced to the phase coupling model [19]. The delay
time could affect the coupled plasmodial oscillator system, because the time during which the state information in one oscillator propagates to the other oscillator is not zero. The time could correspond to the duration of the wave propagation between the two plasmodial oscillators, which can be measured as the increase in wave propagation of the thickness in the transmitted image of the tube
structure. The delay time can be estimated from the wave propagation velocity
(Figure 1.2-3 C) and the channel length L. We found that the delay time depends
mainly on L (strictly, the delay time depends mainly on L but also on W [18]).
Indeed, the delay time was estimated as 26 s, when $L=10$ mm and $W=500$ μm,
which cannot be neglected against one period of oscillation (~ 80 s).

Furthermore, we investigated the time delay effect in the plasmodial oscillator
system by comparing the experimental results with the theoretical results of the

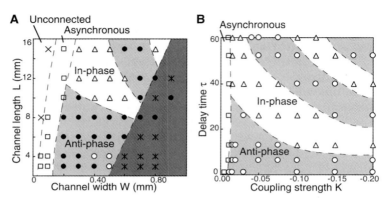

Figure 1.2-6. Phase diagrams. (**A**) Experimental result with the parameters W and L. Crosses mean that the tube was not formed between two oscillators, squares mean that two oscillators are asynchronous, filled circles, open circles, triangles and asterisks show quasi-periodic anti-phase, anti-phase, in-phase, and complicated oscillation, respectively. The data in the shaded region with dark grey were not considered, since the dendritic tube structures are formed and complicated oscillation patterns are observed. (**B**) Numerical result with the parameters, coupling strength and delay time, obtained from model equations of a phase-coupled oscillator system. Squares, circles and triangles show asynchronous, anti-phase, and in-phase oscillation, respectively. (Adapted from [18])

delayed oscillator system. Figure 1.2-6 shows the phase diagram with two parameters W and L for the experimental result (Figure 1.2-6 A) and with the coupling strength K and the delay time τ for the theoretical result (Figure 1.2-6 B). In both diagrams, three major features appear, ie, asynchronous, anti-phase, and in-phase oscillations, depending on the parameters. Interestingly, the transitions from/to anti-phase to/from in-phase were repeated as the parameters W or K and L or τ increased. By further analysis with the common frequencies, we first showed the existence of the time delay effect in the biological coupled oscillator system [18].

1.2.4   Multiple Oscillator System

A practical biological system seems to consist of more than two oscillator units. This would lead to a richer behavior. For further observation and analysis of such systems, we constructed multiple oscillator systems of the plasmodium.

Figure 1.2-7 A shows the microfabricated structure for multiple oscillator systems that consist of three, four, five and more oscillators and couplings (Figure 1.2-7 B). For these structures, the cross-shaped structures in the wells (Figure 1.2-7 C) must to be fabricated to support the isolated sandbank structures (Figure 1.2-7 A) surrounded by multiple channels. The microfabricated structures

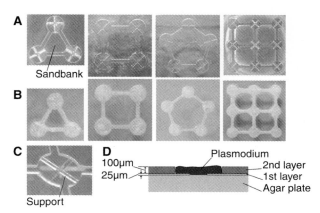

Figure 1.2-7. Multiple-oscillator system. (**A**) Microfabricated structures for three, four, five, and multiple coupled oscillators. (**B**) The plasmodium patterned by the microfabricated structures. (**C**) Magnified picture of the support structure in the oscillator part. (**D**) Cross-sectional view of the system.

comprise two layers; the first thin layer is for the support structures (the cross-shaped structures) and the second thick layer is for patterning of the plasmodium (Figure 1.2-7 C and D). The first layer is about 25 μm thick; we have confirmed that the layer does not perturb the oscillation of the plasmodium in the well.

Generally, it is difficult to analyze a multiple-oscillator system mathematically. However, some theorists have obtained an approximate result by restricting the system to geometrically symmetric systems. Without a knowledge of the precise mechanism of oscillators, the symmetric Hopf bifurcation theory based on group theory provides an elegant mathematical understanding of spatio-temporal oscillation pattern formations [20]. We therefore constructed symmetrically coupled oscillator systems in a ring as shown in Figure 1.2-7 A and B. We observed most of the oscillation patterns predicted by the theory and found that even the biological system, where its intrinsic mechanism is not always known, follows the simple mathematical model [21]. This approach could help us to understand the complicated behavior of living beings.

1.2.5 Integrated Observation System with a Temperature Control Device

As a further step, it would be interesting to investigate the response of a biological system to external stimuli to understand its information processing mechanism. We are developing an integrated observation system with a temperature control device. The plasmodium responds to a temperature stimulus and shows

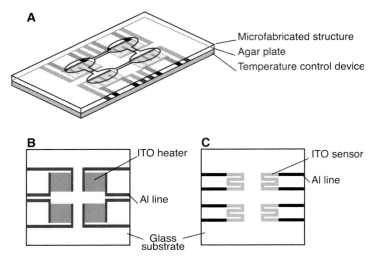

Figure 1.2-8. (**A**) Integrated observation system with temperature controller. (**B**) First layer with ITO heater. (**C**) Second layer with temperature sensor made of ITO. (Partly adapted from [24])

escaping/gathering behavior from/to a heat source with lower/higher temperature [22, 23]. It is important to observe the oscillatory phenomena in the plasmodial oscillator system simultaneously applying the local temperature stimuli. Thus, we designed the temperature control device using transparent resistive material, indium tin oxide (ITO), as heaters and sensors [24]. Figure 1.2-8 A shows a schematic diagram of the integrated observation system with the temperature control device that consists of a microfabricated structure for patterning of the plasmodium and a device for applying a temperature stimulus. The temperature control device comprises two layers. In the first layer, the ITO heater is patterned on a glass substrate (Figure 1.2-8 B) and in the second layer the ITO sensor is patterned on the first layer (Figure 1.2-8 C). The SiO_2 layers are deposited between the two layers and on the second layer for electrical isolation. Aluminium is patterned as the other electric lines. The multiple combinations of heaters and sensors enable us to apply multiple stimuli to arbitrary oscillators of the plasmodium. This is important for the observation of the behavior of living beings in their complicated environment. In addition, the advantage of this system is the transparency of the stimulus device, because it would be generally essential to observe cells, eg, a neuronal system, optically [25]. This system provides us with a method of simultaneous observation with stimulation.

1.2.6 Conclusion

We applied the microfabrication technique to realize a cell-sized structure. Furthermore, for the investigation of the plasmodial slime mold system, we introduced the mathematical concept of nonlinear dynamics to perform side-by-side analysis with the experimental results and the theoretical predictions. These approaches could be generally applicable to the other complicated biological systems, especially collectives of cells such as neurons, cardiac cells, liver cells, microorganisms, etc. For such analyses, the microfabrication technique must be one of the key technologies to construct a simple but essential observation system for a complicated biological system.

1.2.7 Acknowledgments

The authors thank Dr. I. Endo, chief scientist of the Biochemical Systems Laboratory at the RIKEN Institute, Japan, where part of this work was performed. The authors also thank Dr. K. Hosokawa, AIST, Japan, for his helpful advice on the microfabrication of SU-8, and Dr. T. Yamamoto, of the University of Tokyo, Japan, for his help with the fabrication of the ITO heater/sensor. Dr. R. Tanaka, Keio University, Japan, and Dr. H. Yamada and Professor T. Nakagaki, Hokkaido University, Japan, are collaborators in the analysis of oscillation patterns in multiple oscillator systems, and the authors are grateful to them for fruitful discussions.

1.2.8 References

[1] Winfree, A. T., *The Geometry of Biological Time*; New York: Springer, 1980.
[2] Rapp, P. E., *J. Exp. Biol.* **81** (1979) 281–306.
[3] Kuramoto, Y., *Prog. Theor. Phys.* **71** (1984) 1182–1196.
[4] Strogatz, S. H., *Nonlinear Dynamics and Chaos*; Reading, MA: Addison-Wesley, 1994.
[5] Jimbo, Y., Robinson, P. C., Kawana, A., *IEEE Trans. Biomed. Eng.* **40** (1993) 804–810.
[6] Matsuzawa, M., Krauthamer, V., Potember, R. S., *Johns Hopkins Appl. Tech. Dig.* **20** (1999) 262–270.
[7] Maher, M. P., Pine, J., Wright, J., Tai, Y., *J. Neuros. Methods* **87** (1999) 45–56.
[8] Knowles, D. J. C., Carlile, M. J., *J. Gen. Microbiol.* **108** (1987) 17–25.
[9] Yoshimoto, Y., Sakai, T., Kamiya, N., *Protoplasma* **109** (1981) 159–168.
[10] Yoshimoto, Y., Matsumura, F., Kamiya, N., *Cell Motility* **1** (1981) 433–443.

[11] Yoshimoto, Y., Kamiya, N., *Protoplasma* **95** (1978) 89–99.
[12] Baranowski, Z., *Acta Protozool.* **17** (1978) 377–388.
[13] Ueda, T., Matsumoto, K., Akitaya, T., Kobatake, Y., *Exp. Cell. Res.* **162** (1986) 486–494.
[14] Kamiya, N., *Cytologia* **15** (1950) 194–204.
[15] Takamatsu, A., Fujii, T., Yokota, H., Hosokawa, K., Higuch, T., Endo, I., *Protoplasma* **210** (2000) 164–171.
[16] Takamatsu, A., Fujii, T., Endo, I., *BioSystems* **55** (2000) 33–38.
[17] Ueda, T., Mori, Y., Nakagaki, T., Kobatake, Y., *Photochem. Photobiol.* **48** (1988) 705–709.
[18] Takamatsu, A., Fujii, T., Endo, I., *Phys. Rev. Lett.* **85** (2000) 2026–2029.
[19] Schuster, H.G., Wagner, P., *Prog. Theor. Phys.* **81** (1989) 939–945.
[20] Golubitsky, M., Stewart, I., in: *Multiparameter Bifurcation Theory, Contemporary Mathematics*, Golubitsky, M., Guckenheimer, J. (eds.); Providence, RI: Ams, 1986, Vol. 56, p. 131.
[21] Takamatsu, A., Tanaka, R., Yamada, H., Nakagaki, T., Fujii, T., Endo, I., *Phys. Rev. Lett.* **87** (2001) 078102.
[22] Matsumoto, K., Ueda, T., Kobatake, Y., *J. Theor. Biol.* **122** (1986) 339–345.
[23] Matsumoto, K., Ueda, T., Kobatake, Y., *J. Theor. Biol.* **131** (1988) 175–182.
[24] Yamamoto, T., Fujii, T., Nojima, T., Hong, J., Endo, I., *Proc. SPIE* **4177** (2000) 72–79.
[25] Shugihara, H., in: *Sensors Update*, Baltes, H., Gopel, W., Hesse, J. (eds.); Weinheim: Wiley-VCH, 1999, Vol. 6, p. 243.

List of Symbols and Abbreviations

Symbol	Designation
K	coupling strength
L	channel length
W	channel width
τ	delay time

Abbreviation	Explanation
CCD	charge coupled device

1.3 Towards MEMS Probes for Intracellular Recording

Y. HANEIN, K. F. BÖHRINGER, R. C. WYETH, and A. O. D. WILLOWS,
University of Washington, Seattle, WA, USA

Abstract

Simultaneous, multi-site recording from the brain of freely behaving animals will allow neuroscientists to correlate neuronal activity with external stimulation and behavior. This information is critical for understanding the complex interactions of brain cells. Recent interest in microelectromechanical systems (MEMS) and in particular in bio-MEMS research has led to miniaturization of microelectrodes for extracellular neuronal recording. MEMS technology offers a unique opportunity to build compact, integrated sensors well suited for multi-site recording from freely behaving animals. These devices have the combined capabilities of silicon-integrated circuit processing and thin-film microelectrode sensing. MEMS probes for intracellular recording may offer significantly improved signal quality. Here we discuss the basic concepts that underlie the construction of intracellular MEMS probes. We first review the basics of neuronal signaling and recording, and the principles of microelectrode technology and techniques. Progress in MEMS technology for neuronal recording is then discussed. Finally, we describe MEMS probes for intracellular recording, viz., fabrication of micro-machined silicon needles capable of penetrating cell membranes. Using these needles, we recorded localized extracellular signals from the hawk moth *Manduca sexta* and obtained first recordings with silicon-based micro-probes from the *inside* of neurons, using an isolated brain of the sea slug *Tritonia diomedea*.

Keywords: Microelectrodes; Intracellular; Bio-fouling; Microelectrode arrays; Silicon probes; Neurons

Contents

1.3.1   Introduction

Understanding brain structure and function has challenged mankind for many centuries. Only during the second half of the nineteenth century, due to advances in scientific tools and in particular the development of a special staining techniques, did scientists reach the understanding that the brain consists of a complex, interactive network of single cells (neurons). The number of cells varies from $\sim 10^{11}$ in humans down to a few hundred in small invertebrates such as leeches. These networks interact and enable living creatures to function, decide, learn, remember, and achieve consciousness. In recent decades, with the rapid development of neuroscience techniques, researchers have obtained detailed information about the function and organization of the brain, and the structure (Figure 1.3-1) and operation of neurons.

Modern tools to investigate the brain consist of a wide variety of techniques such as positron emission tomography (PET) scanning, functional magnetic resonance imaging (fMRI), and electrical recording, among others. Each of these tools is optimized for specific applications. Different techniques may complement each other and several techniques may be used simultaneously. Existing techniques fall into many categories, such as invasive versus noninvasive, local versus regional, and therapeutic versus research. Clearly, for diagnosis purposes a

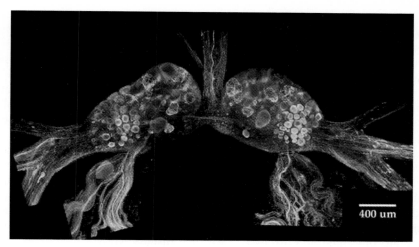

Figure 1.3-1. An immunolabelled brain of *Tritonia diomedea* (sea slug). Different dyes are used to mark cells that contain different neurotransmitters, specific chemicals used by neurons to communicate with each other. It is important to note that only a fraction of the cells are stained and so the stained cells appear on an unstained background, that is actually filled with other cells. Image by Jim Beck.

noninvasive technique is preferred and a system such as PET scanning, a noninvasive imaging procedure that visualizes local changes in the cerebral blood flow, has established itself as a standard diagnosis tool. Despite the wide variety of existing tools and despite the great progress of the recent years, many questions related to brain activity are still unresolved and principles that underlie many brain processes are still unknown. It is widely acknowledged that further progress will be enhanced by novel probing tools.

Neuroscience has accomplished much in understanding how individual neurons function and how they work together in small populations. However, much less is known about how different parts of the nervous system are integrated. For example, how are the different senses used to navigate through the environment? This type of question exceeds the capabilities of current tools. Conventional tools require extensive laboratory apparatus, and eliminate the normal sensory world. In order to understand nervous system function at the system level, we need to reduce the influence of laboratory apparatus. An unencumbered setting allows animals to have real sensations and feedback from their own activities.

A better systems-level recording tool will allow researchers to perform high-fidelity multi-site intracellular recording from freely behaving animals. This will allow neuroscientists to study the correlation between (a) neuronal activity and external stimulation and (b) neuronal activity and behavior, and also to understand better the communication patterns inside neural networks. As we will show below, existing intracellular probes are well suited for studying the physiology and the processes of *single, living* cells, in particular for understanding heterogeneous living cell populations that make up dynamic systems such as neural networks. However, despite their obvious advantages, existing intracellular probes are usually large, bulky, fragile, and not available in arrays.

Microelectromechanical systems (MEMS) technology may provide an opportunity to develop multi-site recording from freely behaving animals. MEMS offer small overall dimensions, easily prepared arrays, and built-in integrated circuit capabilities [1]. Recently, researchers have demonstrated the capabilities of planar microelectrode array (MEA) probes to study specific neurological problems [2, 3].

This chapter reviews the basic principles of intracellular recording and describes the motivation and challenges associated with the fabrication of intracellular MEMS probes. We begin our discussion with a short review of neuronal structure and activity. We then discuss neuronal signals and how these signals are recorded. We dwell on the properties of different recording elements and in particular on microelectrodes. The differences between extra- and intracellular recording are presented and discussed. We then discuss MEMS electrodes and the techniques used in their construction. Finally, we present the details of MEMS-based intracellular probes that we have constructed recently. Using these needles, we now obtain extremely localized extracellular signals and have made first recordings with silicon-based micro-probes from the *inside* of neurons.

1.3.2   Principles of Neuronal Recording

Neuronal communication is at the core of brain activity, and understanding its signaling is the key to understanding how the brain works. Many neuroscientists therefore wish to be able to record these signals in real time and from large numbers of neurons simultaneously. To understand neuronal recording we first need to clarify some of the basic conceptual and structural issues related to the brain. A good starting point is understanding individual electrically excitable brain cells (ie, neurons) and their connectivity. A comprehensive description of these processes can be found elsewhere [4–7]. Here we briefly summarize the basic principles.

1.3.2.1   Neurons

Neurons consist of a cell body, axons, and dendrites all enclosed by a thin, fragile, phospholipid bilayer membrane (Figure 1.3-2). The signals generated by neurons are transmitted by their axons to synapses at their terminals. These terminals usually contact dendrites or the cell body of another neuron (post-synaptic neuron). The inputs to a neuron are often delivered by chemicals (neurotransmitters) diffusing across one or many synapses from presynaptic neurons.

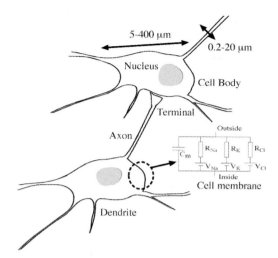

Figure 1.3-2. Neuron structure. Dendrites and the cell body receive 'inputs' across synapses, and the axon sends signals on to other neurons across synapses at its terminals. The cell membrane is a thin insulating layer that separates two conducting layers (the interior and the exterior of the cell). Therefore, the membrane can be regarded as a capacitor with additional potential sources and resistors representing the different ionic channels in the membrane. A simplified electrical equivalent circuit of the membrane is shown in the inset following the well-known Hodgkin-Huxley approach. According to this model a neuron can be represented by an electrical equivalent circuit that includes elements representing different membrane components. Many of the mechanisms underlying these components are now well understood.

The chemical signals are transduced into electrical signals, which are regenerated actively along the cell membrane. These signals, whether arising on dendrites or the cell body, may culminate in another regenerative electrical membrane impulse, transmitted along the axon to the terminals. Again, the impulse causes release of neurotransmitters that diffuse across the synapse to the next neuron, and the cycle repeats. Direct electrical communication not mediated by chemicals is also common between neurons. In either case, the impulse traffic and underlying synaptic potentials are central interests of neuroscience. To understand measurement of this bio-electricity we need to proceed one step further into the structure of the cell.

1.3.2.2   Neuronal Signaling

Neurons are surrounded by a semi-permeable membrane. The membrane insulates the conducting interior from fluids surrounding the cell. Ionic pumps in the membrane transport ions from lower to higher electrochemical potential, maintaining electrochemical gradients across the cell membrane. Owing to ionic concentration gradients and differential membrane permeability, primarily to K^+, Na^+ and Cl^-, neurons sustain a DC potential (resting potential) across the membrane. This resting membrane potential (V_m) varies for different systems, with typical values of –90 mV for humans, –70 mV for the squid giant axon, and –50 mV for sea slug neurons.

In neurons, the membrane potential is modulated by ionic currents through several different types of ion-specific channels across the cell membrane. These channels can selectively permit different ions to cross the membrane (down their electrochemical gradients). Depending on their timing, location, and ion specificity, ionic movements alter the membrane potential with corresponding rates, amplitudes, and direction.

There are two broad categories of electrical events brought about by these ionic movements: passive and active. A very common example of a passive electrical signal is the post-synaptic potential. Channels on the post-synaptic neuron open in response to the chemical diffusing across the synapse and cause the membrane potential near that synapse to increase or decrease. This voltage change moves *passively* along the membrane, affecting closer areas of membrane more than those more distant. If enough of these small post-synaptic potentials combine in an area of voltage-sensitive channels, then an *active* or regenerative electrical impulse will occur. An active electrical impulse will move along the membrane whenever a changing membrane potential in one area of membrane induces a similar change in membrane potential in an adjacent area. This action potential is a fast and relatively large voltage change, and by actively regenerating itself can travel much farther than passive electrical signals. By propagating along contiguous areas of active membrane, the signal can quickly travel the length of an axon and initiate synaptic transmission to the next neuron. To understand how neurons work, and work together, it is very helpful to record both the small post-synaptic potentials and the resultant transmitted action potentials.

An example of how this process occurs is odor detection [7]. Sensory neuron terminals inside the nasal cavities have extensions with embedded ionic channels. These channels open in response to only a few odor chemicals, creating small 'receptor' potentials, almost identical with post-synaptic potentials. If enough of the specific odor chemical interacts with the sensory neuron terminals, then those small potentials combine additively to reach a certain threshold voltage, triggering a regenerative action potential in the sensory neuron. Different sensory neurons have different types of receptor channels, and thus sensitivity to different chemicals. The combination of receptor potentials and action potentials is used to code the presence or absence of a certain chemical, and the pattern of firing across the whole population of sensory neurons is used to code the odor composition of the air breathed by the animal. Action potentials generated in sensory neurons are transmitted to the brain, where the olfactory information is processed. Perhaps, if the odor is noxious, through a series of synaptic potentials and action potentials, the information will be transmitted to parts of the brain which control movement, allowing the animal to turn its head away from the odor source.

How do neuroscientists observe and understand this process? Fundamentally, there are two ways to record the transmembrane voltages associated with neuron activity. The intracellular method uses two electrodes: one inside the cell (intra-cellular electrode), and the other outside (reference electrode). This method re-cords transmembrane signals, ie, post-synaptic potentials and the action poten-tials (impulses). The other, extracellular method, places two electrodes outside the cell, one very close to the cell and one further away. Although the signals re-corded in this way do not strictly measure the transmembrane currents generated by the neuron, any regenerative or other large currents generated nearby the elec-trodes may cause voltage differences between the two electrodes. Extracellular recording is therefore effective for recording action potentials, but is less likely to detect post-synaptic potentials or other small potential changes. All changes in membrane potential hold information about the way neurons integrate their in-puts and communicate their outputs, and by recording and understanding these signals one can observe the basics of brain communication. Let us now turn to the details of the tools one can use to record these signals.

1.3.2.3   Neuronal Recording

Over the past five decades, a wide variety of electrodes have been developed to record bio-electric events. A subclass of these electrodes are small, localized probes typically used to study neuronal signaling [6]. As we will show below, the electrical properties of a microelectrode determine its reliability as a record-ing transducer. Aspects of design, materials and fabrication may cause distortion, electrical noise, and instability.

Most electrodes are either metallic or glass micro-pipettes [8]. We begin this section with a review of these two most common recording devices and then dis-cuss their two associated alternative recording approaches, viz., intra- and extra-cellular recording. This discussion will point out the relevant issues in designing a silicon-based probe suited for intracellular recording. Finally, we will introduce planar MEAs.

A straightforward approach to realizing a small, localized probe is to use an exposed tip of a sharp, insulated, conducting wire (Figure 1.3-3b) [6]. In these probes the signal is transferred from the tip through the wire while a dielectric material provides the insulation between the wire and the surrounding environ-ment. An additional standard technique to achieve tips with sub-micrometer di-mension is by pulling heated glass capillaries [4] (Figures 1.3-3c and 1.3-4). These pulled micro-capillaries can be easily transformed into microelectrodes by filling them with an electrolyte (typically KCl or KOAc) and placing a Ag/AgCl electrode in the electrolyte. The glass wall (Figure 1.3-4b) provides ionic insula-tion and ensures virtually no leak current.

Despite a number of significant differences between the metal and micro-pi-pette electrodes, the underlying principles are similar and have been investigated extensively. In essence, all recording electrodes consist of a metal-electrolyte in-terface. Unlike the simple ohmic metal-metal contact, a metal-electrolyte contact is a rather complex system [9].

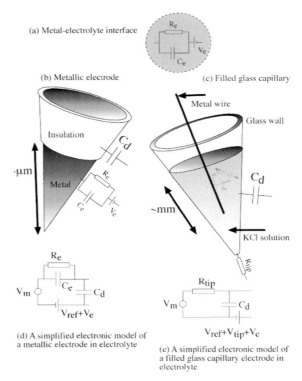

(a) Metal-electrolyte interface

(b) Metallic electrode

(c) Filled glass capillary

Insulation

Metal

Metal wire

Glass wall

~µm

~mm

KCl solution

R_e

V_m C_e C_d

$V_{ref}+V_e$

R_{tip}

V_m C_d

$V_{ref}+V_{tip}+V_e$

(d) A simplified electronic model of a metallic electrode in electrolyte

(e) A simplified electronic model of a filled glass capillary electrode in electrolyte

Figure 1.3-3. Bioelectricity-recording electrodes. (**a**) A simplified electronic model for a metal-electrolyte interface. V_e represents the potential drop across the interface, R_e the resistance and C_e the capacitance of the metal-electrolyte interface. This model is used for the metallic and the filled-glass microelectrodes. (**b**) A schematic drawing of a metallic electrode. (**c**) A schematic drawing of a filled glass capillary. (**d**) A simplified electronic model for a metallic electrode in an electrolyte solution. (**e**) A simplified electronic model for a filled-glass capillary electrode in an electrolyte solution. Comparison between the circuits in (**d**) and (**e**) demonstrates the effect of the geometry of the electrode on its electrical properties. In metallic microelectrodes the metal-electrolyte interface is located at the recording tip and therefore accounts for almost all the impedance of the electrode.

Various chemical reactions may take place when a metal is introduced into an electrolyte. These reactions may involve dissolution of the metal in the case of partially soluble metals, or electron exchange between the metal and the solution as in the case of noble metals. The result of these chemical reactions is the formation of an equilibrium charge gradient at the interface (usually referred to as the electric double layer) which is accompanied by a buildup of an electric potential across the interface. The details of the space charge layer are predicted by theoretical models (Helmholtz, 1897), which can then be used to analyze the electrical properties of an electrode.

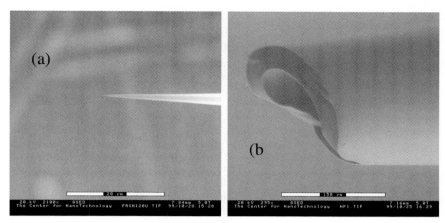

Figure 1.3-4. Glass micropipettes. (**a**) An Environmental scanning electron microscope (ESEM) image of the tip of a pulled glass capillary. The scale bar is 20 μm. Similar tips are commonly used for intracellular neuronal recording. (**b**) An ESEM image of a broken glass capillary. The glass provides superb insulation between the inner and the outer sides of the probe. The scale bar is 150 μm.

Owing to the capacitive nature of the electric double layer, a metal-electrolyte interface is, in fact, an electrolytic capacitor (eg, for platinum $C_e = 20$ μF/cm^2 at 1 kHz). In addition to the capacitive nature of the metal-electrolyte interface, we should also consider its resistive nature. A metallic electrode and an electrolyte maintain equilibrium potential and a balance between influx and efflux currents of electrons (or ions). By applying an external potential, the equilibrium current is unbalanced, and the induced current can then be expressed [10] by

$$i = i_0 e^{F\Delta V/2RT} - i_0 e^{-F\Delta V/2RT} \tag{1}$$

with i_0 being the exchange current density (values may range from pA/cm^2 up to 10 A/cm^2), ΔV is the applied potential, F the Faraday constant, R is the gas constant, and T is the temperature. For small voltages, Equation (1) can be linearized and expressed as

$$i = i_0 F/RT\Delta V . \tag{2}$$

At room temperature the electrode resistance (typically denoted as the charge transfer resistance) can be expressed as $R_e = \Delta V/i = 0.06/i_0$. For platinum, $i_0 = 4.5 \times 10^{-6}$ A/cm^2, which corresponds to $R_e = 1.3 \times 10^{12}$ Ω/cm^2. Clearly, for very small electrodes the above expression results in very significant resistances, which are in many cases one of the major complications in the construction of metallic electrodes.

In practical terms, the space charge layer at the metal-electrolyte interface can be simply modeled as a voltage source (V_e) in series with a capacitor (C_e) and a resistor (R_e) in parallel (see Figure 1.3-3 a) (this model is appropriate for low frequencies; at high frequencies the impedance can be modeled by an equivalent circuit of resistor and capacitor in series [10]). For many materials, the values of the simplified components have been determined [10, 11]. It is important to note that these elements cannot be treated as a capacitor or resistor with fixed values. In fact, the values of these elements vary with frequency [10, 12, 13] and also with material, electrolyte, and temperature. This dependence reflects changes in the double-layer properties with these parameters.

Let us begin with analyzing the properties of metallic recording electrodes (for detailed explanations, see [11]). It should be noted that the intracellular MEMS probes, which we fabricate, are very similar to these probes. In Figure 1.3-3 b we draw a schematic presentation of an electrode. We also illustrate the major electrical components. The electrode consists of two components: a metallic tip and an insulated shank. The metallic tip can be represented (when placed in an electrolytic solution) by a resistor (R_e), a capacitor (C_e), and a potential source (V_e). The insulation of the electrode shank separates the metallic conductor of the electrode from the conducting electrolytic solution and therefore can be simply represented (when placed in electrolytic solution) as a capacitor (C_d). Additional components in the system are the shank and electrolyte resistances (determined by the geometric surface area of the tip [10, 12]). The equivalent electric circuit of such a probe in recording conditions (with a reference electrode) is shown in Figure 1.3-3 d and follows directly from the model in Figure 1.3-3 a. Note that some elements, such as the resistance of the electrode shank, and the intra- and extracellular liquid resistances were omitted as they are negligible in comparison with the other elements in the circuit.

The equivalent electrical circuit for the micro-pipette electrodes is shown in Figure 1.3-3 e. Here, too, we can neglect various elements. In this example the metal-electrolyte resistance and capacitance (R_e and C_e) are negligible owing to the large surface area of the contact between the electrolyte and the wire. The two major components that determine the electrode impedance are the electrode resistance (R_{tip}) and the glass wall capacitance (C_d). The first is the resistance of the electrolyte through the narrow tip opening, and is determined by the tip diameter. The capacitance is determined by the glass wall thickness.

If we compare Figures 1.3-3 d and 1.3-3 e it becomes apparent that the geometric differences between the two electrodes produce significant electrical differences. In metallic microelectrodes the metal-electrolyte interface is located at the recording tip and therefore accounts for almost all the impedance of the electrode. The electrical properties of a micro-pipette are dominated by the glass tip resistance. Typically, glass micro-pipettes have DC resistances in the order of 10–200 MΩ, while metallic electrodes may have DC resistance larger by at least two orders of magnitude. On the other hand, in AC, metallic electrodes outperform micro-pipettes which may poorly represent rapidly changing signals owing to their large shunt capacitance. These differences can have a direct impact on the performance of the electrodes for different applications [13]. For example, the electrolyte-filled electrodes will perform as low-pass filters. Therefore, they are most suited for intracellular recording of relatively slowly changing potentials. On the other hand, the metallic electrodes act as high-pass filters, which suggests their use for more rapidly changing signals. The impedance of the electrode (determined by the materials and geometry) and the impedance of the external path determine the amount of distortion of the recorded signal [13, 14]. Careful choice of metallic electrode parameters can significantly improve their DC performances, their stability, and their distortion.

By considering the differences between metallic electrodes and micro-pipettes, we can explain the parameters and considerations which are related to the design of recording electrodes. Let us now turn to explain when and how these parameters become relevant in an experiment. The performance of an electrode is largely determined by the event under investigation (ie, rapidly or slowly changing signals) and by the anatomical location. Recordings can be made with the electrode inside or outside the cell. Clearly, placing electrodes inside the cell imposes some major challenges on the construction and handling of the electrodes. However, there are several major drawbacks to extracellular recording that make intracellular recording worthwhile. First, information gathered by extracellular sensing may not be exclusive to a single cell (this statement is valid for the case of poor sealing between the electrode and the cell membrane; see the discussion below). Rather, it may be an average over several cells located at the vicinity of the probe. Second, extracellular probing does not provide critical information about DC conditions or slowly changing potentials across the cell membrane. Only the time of occurrence of action potentials can be recorded, not the details of their form. This is a direct result of the capacitive nature of the cell membrane (typical values of the order of 1 $\mu F/cm^2$).

To understand these differences better, we present in Figure 1.3-5 a comparison between extra- and intracellular recording results. The recording was from an isolated brain of *Tritonia diomedea*. *Tritonia* is a marine mollusk indigenous to the Pacific Northwest; its hallmarks are extraordinarily large brain nerve cells (see Figure 1.3-1), identifiable sensory and motor functions associated with these brain cells, and robust response to surgical insult. Brain preparation methods for this animal allow recording and stimulating a brain during voluntary and reflexive movement [15].

Two electrodes were used to record simultaneously from the same neuron. The first electrode was an intracellular electrode inserted into the neuron. The second extracellular electrode was placed directly adjacent to the first electrode and suction was applied to seal the electrode against the edge of the brain, directly over the cell. Although both records show the action potentials, the dynamic range of the intracellular recording is roughly three decades larger than that of the extracellular recording. Missing from the extracellular record are both the action potential shape and DC changes in the resting potential (here induced by positive and negative current injection). With higher amplification, small synaptic potentials which are clearly visible in the intracellular record, are invisible in the extracellular record. In an extracellular recording, a nearby-firing neuron can confound and sometimes obscure action potentials from the cell of interest; this effect never occurs in intracellular recordings.

To conclude our discussion so far, intracellular recording is important when one wants to know more than just when a neuron fired (and even then, extracel-

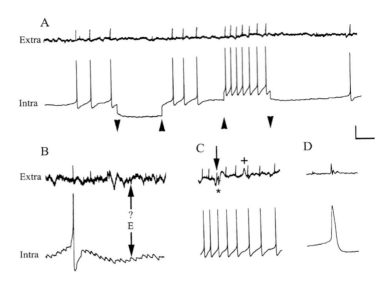

Figure 1.3-5. Extracellular and intracellular recordings from a single reidentifiable neuron, in an isolated *Tritonia* brain. An intracellular electrode was inserted into the neuron. An extracellular electrode was applied with suction, directly nearby. (**A**) Although both records show the action potentials, the dynamic range of the intracellular recording is roughly three decades larger than that of the extracellular recording. Missing from the extracellular record are both the action potential shape and the DC changes in the resting potential (here induced by positive and negative current injection (arrowheads), but also occurring *in vivo*). (**B**) With higher amplification, small synaptic potentials (eg, excitatory post-synaptic potentials, one marked by **E**) clearly visible in the intracellular record, are obscured in the extracellular record. (**C**) In an extracellular recording, a nearby-firing neuron can confound and sometimes obscure action potentials from the cell of interest (arrow); this effect is absent from the intracellular record. (**D**) A single action potential shown on an expanded time-scale. Analyzing an extracellular record is clearly more difficult than analyzing an intracellular record. Scale bar: (**A, C**) extracellular 25 µV, 5 s; intracellular 25 mV, 5 s; (**B**) extracellular 10 µV, 5 s; intracellular 10 mV, 5 s; (**D**) extracellular 25 µV, 0.2 s; intracellular 25 mV, 0.2 s.

lular recordings are hard to associate with a single, identifiable neuron). Intracellular recordings are useful to observe the full electrical activity of single neurons: the small DC changes associated with synaptic interactions, the shape of action potentials (which can also be critical), and the timing of action potentials. For these reasons, our goal is to build a silicon-based intracellular electrode. Therefore several requirements are imposed. Special attention has to be directed not only to their geometry but also to the fidelity of the probes under DC and AC conditions. Our silicon-based electrodes should have performances similar to those of the intracellular micro-pipettes. It is very important to emphasize that the extracellular recording we discussed above was performed with a tight seal and therefore represents the best case scenario for extracellular recording.

To summarize, here are some topics that must be considered in the design of silicon-based intracellular probes. Special attention has to be directed to possible instability and noise sources. Stability can be improved by adequate choice of metal. It is widely accepted that silver/silver chloride electrodes are the most stable electrodes [11, 14]. When discussing microelectrodes we should also consider the following two main sources of noise. The first is associated with the unstable metal-electrolyte interface. Here too, the noise level is determined by the metal used. However, this noise may also be dramatically affected by the preparation procedure (ie, the exact parameters of chloriding) as well as the final hookup to the measurement apparatus. A second noise source is thermal noise. Thermal noise effects are related to the resistance of the probe and can be minimized by reducing the electrode impedance. An additional major source for artifacts is the amplification stage. Incompatibility between the probe and the amplifier may distort the signal. Finally, to minimize galvanic potentials, a reference electrode of the same metal as the recording electrode should be used.

1.3.2.4   MEMS Neural Probes

Owing to the multicellular nature of the nervous system, simultaneous recording from a large number of neurons may be helpful. MEMS devices are particularly promising for achieving this goal owing to their small dimensions and the ease with which multi-site devices can be produced. Indeed, extensive effort in the past three decades has shown the potential of planar MEA devices for neurological and electrochemical sensing applications.

What is MEMS and what makes MEMS such an appealing technology for neurological applications? MEMS technology takes advantage of micro-fabrication techniques to construct a wide variety of small electromechanical and also chemical and biological devices [16]. The number of existing techniques, such as metal deposition, bulk etching, dielectric deposition, and molding, is so vast nowadays that the miniaturization of various tools and devices is becoming an everyday reality. MEMS technology has become dominant during recent decades in various applications such as accelerometers, digital mirror displays, and DNA chips.

Silicon-based planar microelectrode arrays were developed with the forethought to allow both *in vivo* and *in vitro* multiple site recording. These devices support the combined capabilities of silicon integrated-circuit processing with thin-film microelectrode sensing. The pioneering work by Wise et al. [18] has been followed by numerous studies that exploited integrated-circuit technology to build neurological microelectrodes. These devices typically consist of metallic electrodes, such as iridium [17], gold [19, 20], and platinum [21], which are photolithographically patterned on passivated silicon substrates (Figure 1.3-6). The interconnects are passivated by a dielectric layer. The *in vivo* designs include a release process that separates needle-shaped devices from the silicon wafer. For a review on the design and realization of thin-film microelectrodes, see [12].

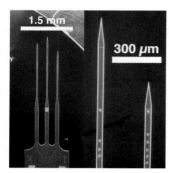

Figure 1.3-6. Silicon-based planar microelectrode arrays (MEAs) for cortical recording. D. Kewley and G. T. A. Kovacs. By permission.

An additional benefit of MEMS devices is the wide variety of additional sensors or effectors that can be integrated with the recording electrodes. With such elements neuronal recording can be linked with chemical stimulation using fluidic channels and valves [22], or temperature control using micro-heaters [23].

Thin-film microelectrodes are produced using standard micro-machining processes. Even though most of these techniques were originally developed for the silicon microelectronics industry, and may include the use of some very harsh chemicals, completed devices made of silicon, noble metals, and dielectric layers, such as silicon dioxide, nitride, or polyimide, are not toxic and can be successfully used to interface with biological elements. A very detailed study by Kristensen et al. [24] demonstrated that the coupling between brain tissues and silicon-based chips had little effect on the tissue under investigation. However, the overall compatibility of the device with the biological environment includes several other factors that have to be considered. (Bio-compatibility is a very common term to describe a proper interface between a biological system and a foreign element. However, this is a very broad and often a very poorly defined concept. To avoid ambiguity, we choose to discuss several specific issues which relate to the interaction between implants and biological systems.)

The first issue is bio-fouling, ie, the strong tendency of proteins and organisms to adsorb physically to synthetic surfaces [25]. Bio-fouling by bacteria is a major source of failure for scores of devices, including macroscopic-scale elements such as metal piping [26]. In the microscopic world of MEMS, bio-fouling is a very challenging issue and adsorbed proteins are known to clog devices with small constrictions, such as bio-capsules [27].

The driving mechanism for bio-fouling in live organisms is protein attachment. This process may affect various devices such as pH [28] and glucose [29] sensors. In these cases, the adsorbed protein layers directly affect the operation of the sensor. Protein layers are also responsible for various biological responses, such as cell attachment and activation [30]. Cell attachment may interfere with the optimal operation of the device by, for example, reducing its life span or increasing its power consumption [31].

Protein and cell attachment to a device surface may trigger the response of the immune system, which in turn may result in inflammation. It is therefore important to consider not only the short-term effects of the biological environment on the device (the effects of protein adsorption on electrode performance are known to occur during periods of hours [28] or days [29]) but also the longer-term effects of the device on the hosting environment. It is important to note that these effects may vary for different applications and biological systems.

Clogging of micro-pipette intracellular electrodes over several hours suggests that protein adsorption may interfere with the recording. In the case of metallic intracellular electrodes, the tips are exposed to protein adsorption. This may affect the recording stability. The components of the device outside the cell are also susceptible to cell attachment. To resolve these problems, surface modification techniques [27, 32] can be integrated with standard MEMS processes and can dramatically reduce protein and cell attachment. It has also been shown that a thin non-fouling coating may provide protection to coated electrodes from protein adsorption and cell attachment without compromising their conductivity [32].

Another major problem related to the interface between artificial devices and biological environments is corrosion. Direct contact between the device and the biological system exposes the surface of the device to corrosive aqueous media. The durability of the device is therefore strongly dependent on the properties of the passivation layer and the quality of the adhesion of the different coatings on the device. Passivation layers used in microelectrode fabrication, such as silicon dioxide, silicon nitride, or polyimide, were originally developed as dielectrics for non-corrosive environments and therefore may perform very poorly (failure after several minutes or hours) in electrolyte solutions. The use of these passivation materials for corrosive environments requires special attention. By studying a large number of common barrier materials as a passivation layer for silicon-based microelectrode devices, Fassbender et al. [33] demonstrated that by a careful choice of material and preparation the corrosion resistance of the passivation layer can be maintained for several months. It was shown that by adequate control of deposition conditions and process cleanliness, effects such as stress, pinholes, and particle inclusion were avoided. In return, effects as buckling and swelling were dramatically reduced and the overall corrosion resistance of the devices was improved.

A major consideration in the design of microelectrodes is their geometric interface with the environment. As was briefly mentioned before, a tight seal between the electrode and the cell membrane is favorable for good extracellular recording. These conditions are very hard to reach with flat electrode designs and a special effort was made, in *in vitro* setups, to improve the sealing by shaping the electrode sites into a cup structure [34]. Action potential simulations based on the equivalent electric circuit of neuron-to-electrode contact show significant signal distortions due to inadequate sealing [35, 36]. Clearly, good sealing is very hard to realize *in vivo* with thin-film microelectrode devices and therefore, despite their many appealing advantages, their recording capabilities are limited to applications where the understanding of synaptic interactions is important.

It is important to note that complete sealing between the recording electrode and the cell membrane may not always be a major problem. This may depend on the level of detail that one requires from the recording and also on the exact experimental setup, ie, whether the recording is performed with brain cells or other nerve cells. MEMS surface electrodes, even without complete sealing, can be used to study brain activity of intact and freely behaving animals [2, 37]. The signal-to-noise level is sufficient to allow spike detection and sorting.

Improvement of the signal-to-noise ratio and higher signal amplitude can be achieved by fabricating three-dimensional tip-shaped electrodes (tips extending from the two-dimensional surface of the wafer). Such designs allow electro-physiological recording from inside a cultured tissue [24, 38]. Campbell et al. [38] used thermo-migration to define p-doped columns in n-doped substrates. A dicing saw was used to define pillars in the p-doped regions. This process allows the formation of tall, electrically insulated pillars. The pillars were sharpened by a chemical etch consisting of 5% hydrofluoric and 95% nitric acid. Gold and platinum were deposited on the tips with a metal foil used as a protection mask for the base of the electrodes. Thiébaud et al. exploited the anisotropic etching characteristics of silicon in KOH to form 47 µm tall tips [39]. The tips were passivated, and then deposited with platinum. As in the planar microelectrode case, the metal is passivated with an additional dielectric layer. A thick photo-resist was patterned and used as a mask to expose the electrode tips.

Owing to the advantages of intracellular recording mentioned above, combined with the favorable properties of MEMS as an enabling technology for neuronal recording, the development of intracellular MEMS probes appears to be a promising approach and is the focus of the current study.

1.3.3   Intracellular Neural MEMS Probes

As was discussed in Section 1.3.2.3, pulled glass capillaries electrodes are most suited for DC recording. MEMS probes based on a similar design, namely hollow capillaries [40], may be ideal for such recording but their realization may be difficult. Needle-like electrodes based on rigid, metallized tips suited for intracellular recording require close attention to their DC properties, but their realization may be more feasible than hollow electrodes. In fact, needle array devices have already been realized and used for extracellular recording [38, 39].

The focus of the current study is the fabrication of micro-machined silicon solid needles suited for intracellular probing. Our main effort has been directed towards the construction of tips capable of penetrating the cell membrane as well as optimization of the electronic properties in DC.

We discuss below the main consideration in the design of these probes. We then describe the fabrication steps including initial characterization and optimization of the different components. To demonstrate the probe performances, we used two biological models: *Manduca sexta* (hawk moth) and *Tritonia diomedea* (a sea slug). Our results indicate that the electrodes act as extremely localized bio-sensors.

1.3.3.1 Tip Design

In Section 1.3.2.3, we discussed the parameters required to model and understand microelectrodes. In this section we discuss the specific design of the MEMS electrodes we produce and estimate the typical values of the different components. The discussion is followed by the fabrication details.

A major challenge in producing probes for intracellular sensing is the tip geometry. Intracellular probes must have extremely sharp tips (sub-micrometer dimensions) and they have to be long (>10 μm). These characteristics are necessary for effective bending and penetration of the flexible cell membrane.

The structure of our electrodes is based on solid silicon needles with a conducting silicon base (Figure 1.3-7 a). The surface of this structure is then coated with a metal that forms the metal-electrolyte interface (Figure 1.3-8 b). A dielectric layer is used to insulate the base of the needle from the electrolyte (Figure 1.3-8 c). The insulation has to cover all parts of the electrode other than the very tip. Connection to the electrode is achieved by wiring an insulated conducting wire to the back of the conducting silicon base. The backside of the electrode and the connection to the wire are finally insulated with a thick encapsulation material.

The geometric structure and the properties of the materials used to construct the needles determine their final performances. Let us estimate the low-frequency values of the different components in this circuit. We begin with the metal-electrolyte interface. The interfacial capacitance (C_{tip}) can be estimated by

$$C_{tip} = C_e A_{tip} .$$ (3)

The charge transfer resistance (R_{tip}) can be estimated by the following simple calculation:

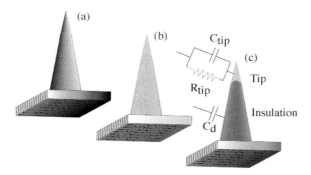

Figure 1.3-7. Schematic drawing of the MEMS intracellular electrodes used in this study. (**a**) The electrodes are made of sharp, bare, silicon needles; (**b**) the bare silicon is then coated with a metal; (**c**) finally, a dielectric passivation layer is deposited on the metal layer and the tip of the metal is exposed. The simplified electric model for the tip and the insulation are also presented. The justification to this simplified model is detailed in the text.

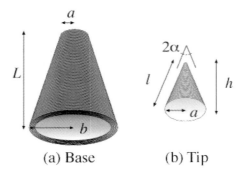

(a) Base   (b) Tip

Figure 1.3-8. Schematic drawing of the MEMS intracellular electrodes. (a) Geometry of the base, (b) geometry of the tip.

$$R_{tip} = R_e / A_{tip} \qquad (4)$$

where C_e and R_e are the specific capacitance and resistance of the metal-electrolyte interface respectively, and A_{tip} is the tip area. In order to maximize the capacitance and minimize the resistance, we need either to increase the effective area of the electrode or to choose a material with high specific capacitance and low charge transfer resistance. A standard way to increase the effective area is by using rough and porous materials, such as porous silicon [41] or platinum black, which can be electroplated on the surface of the electrode [10]. Another alternative to reduce the impedance of the electrodes is to use a material with high specific capacitance such as silver chloride. Silver can be easily deposited by thermal evaporation and chloridation can be achieved by several post-processing means. The details of the chloridation process are discussed below. The estimates below are made for silver/silver chloride electrodes. Noble metal electrodes, such as gold electrodes, would present inferior performances.

For a conical geometry (Figure 1.3-8) the effective area is given by the radius a and the length l:

$$A_{tip} = \pi a l . \qquad (5)$$

For $2a = 9°$ and $h = 20$ µm, $A_{tip} \approx 100$ µm^2. With typical values of $C = 100$ µF/cm^2 (Ag/AgCl), $C_{tip} \approx 100$ pF. For silver chloride electrodes the frequency dependence may be roughly approximated by $C \propto 1/f^{0.4}$, with f being the frequency [11].

To estimate the resistance we can use $R_e = 10^{10}$ Ωµm^2 (for 10 Hz). This value may vary with chloride deposition and can be up to an order of magnitude larger [11]. For $A_{tip} \approx 100$ µm^2, $R \approx 10^8$ Ω. The values used here are for large electrodes, miniaturization of the electrode could affect the effectiveness of the diffusion of the soluble ions close to the electrode tip and may reduce the effectiveness of the electrode.

We now turn to verify the additional components in the circuit. The resistance of the silicon base can be estimated as follows. The resistance of a truncated cone with radii a and b, a length L, and a resistivity ρ is given by

$$R = \rho L / \pi ab \ . \tag{6}$$

For $L = 400$ μm, $a = 1.5$ μm, $b = 35$ μm and $\rho = 0.0045$ Ωcm, $R = 110$ Ω.

The resistance of the metallic coating deposited on the truncated cone can be estimated as follows

$$R = \rho L / 2\pi d \times \ln(b/a)/(b - a) \tag{7}$$

where ρ is the metal resistivity (typical values are on the order of 10^{-6} Ωcm) and d is the metal layer thickness. With $d = 100$ nm, $R = 3\Omega$. The resistance of the metallic coating is comparable to the resistance between the tip and the base. These resistances and also the wiring resistance and the contact resistance is of the same order of magnitude and are low enough to be neglected.

The capacitance of the electrolyte through a passivation layer can be roughly estimated using the parallel plate capacitor equation:

$$C_d = \varepsilon_0 \varepsilon_r A_{\text{Base}} / d_d \tag{8}$$

where ε_0 is the dielectric permittivity of free space, ε_r is the relative dielectric permittivity of the passivation layer (for silicon nitride, $\varepsilon_r = 7.5$), and A_{Base} and d_d are the area and the thickness of the passivation layer, respectively. For $A_{\text{Base}} = 0.25$ mm^2 and $d_d = 50$ nm, $C = 250$ pF. In the present design the capacitance of the needle shank is negligible compared with the capacitance of the electrode base and can be neglected. In this example, the shunt capacitance may impose undesired perturbations on the proper function of the electrode. To minimize these effects, it is important to lower the surface area and to increase the thickness of the passivation.

To conclude, the needles can be modeled by the circuit in Figure 1.3-3 d. The values for R_{tip}, C_{tip} and C_d can be estimated as 10^8 Ω, 100 pF and 250 pF, respectively. This simple calculation gives a possible range of parameters for metallic electrodes for our current design. Further improvement of the design and the preparation may improve these values. It is also important to consider these values with respect to the choice of other components in the circuit, in particular the choice of the amplifier. The value for R_{tip} is comparable to the typical values of glass capillaries and should allow reliable DC recording.

1.3.3.2 Tip Fabrication

To fabricate needles suitable for intracellular recording, we used highly conducting (n-type), 800 μm thick silicon wafers. Similar to the process in [38] we used a dicing saw to dice the wafer in two perpendicular directions to create arrays of tall pillars ($70 \times 70 \times 350$ μm). To sharpen the tips we used reactive ion etching (RIE) with SF$_6$. This is a robust, self-sharpening process, which we optimized in order to obtain long tips with a high aspect ratio. The process requires approximately 45 min and results in sharp, high aspect ratio needles (Figure 1.3-9).

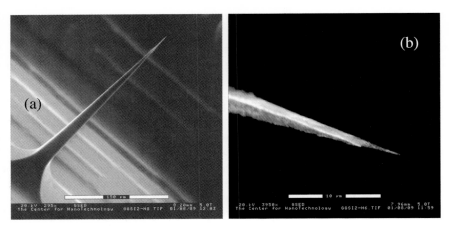

Figure 1.3-9. (**a**) A single needle after the RIE sharpening process. The scale bar is 150 μm. (**b**) A tip of a sharpened silicon needle. The needle is tilted approximately 45° with respect to the plane of the image. The scale bar is 10 μm.

Figure 1.3-9 b shows the tip of a silicon needle that we produced with the process mentioned above. Our process yields a probe geometry that is similar to that of the pulled glass electrodes commonly used in intracellular recording schemes (see Figure 1.3-4 for comparison).

To produce separated needles, the wafer was bonded with crystal bond (or photoresist) to another substrate and cuts to separate the parts were made prior to the sharpening process (Figure 1.3-10). These cuts are approximately 500 μm apart. After the sharpening, the wafer was sputtered with Cr/Au (~ 70 nm) and with silicon nitride (~ 130 nm). Later, using an RIE SF_6 process, the nitride

Separable needle

Cuts used to
separate the dies

Figure 1.3-10. (**a**) An optical microscope image of an array of diced electrodes. The electrodes are glued to a substrate with an adhesive (photoresist). By soaking the sample in acetone the dies can be separated and used as individual probes. (**b**) A schematic drawing of the electrodes after the sharpening process is completed.

layer was slightly etched in order to expose the needle tips. Finally, we soaked the wafer in acetone and released the single needles.

1.3.3.3 Metallization

The main motivation in performing intracellular recording is to be able to measure slowly varying signals. Thus, intracellular recording is reliable and advantageous only if the probe provides a stable recording. The use of metallic electrodes for DC recording is very challenging. One of the main tasks of our study was to investigate the performance of the electrodes and to explore metallizations best suited for DC operation. We investigated two metallization procedures: gold and silver chloride electrodes. The deposition of gold and silver is easily achieved by thermal evaporation with a thin layer of Cr acting as an adhesion promoter.

Gold, as a noble metal, ensures minimal solubility but may result in very high DC resistance and unstable recording. Ag/AgCl electrodes are commonly used to ensure high stability in physiological probing including in MEMS devices [42]. It was suggested that even the reliable Ag/AgCl electrode may fail to support very high fidelity recording. This may be due to interactions between the silver and organic molecules or to the effect related to miniaturization of Ag/AgCl electrodes [13]. Our results, discussed below, show a dramatic improvement of electrode stability and resistance by using Ag/AgCl electrodes.

1.3.3.4 Silver-electrode Chloriding

The performances of silver chloride electrodes depend very strongly on the preparation process. A rigorous review of the preparation and properties of silver chloride electrodes can be found in [11].

To test silver chloride electrodes for DC recording, we investigated the properties of sharp (under 1 μm tip dimensions), silver-coated electrodes in terms of resistance and time constant, before and after chloriding. The electrodes were separated from the holding substrate and individual electrodes were wired and tested. An electrolyte solution droplet was generated at the tip of a syringe and the tips were immersed in the droplet. The surface tension of the droplet allows control over the length of the tip immersed (20–100 μm of the tip is estimated to be immersed in the solution).

The tips were dipped in solution and the DC resistance without chloride was initially 10–20 MΩ (corresponding to $\rho = 1-10 \times 10^9$ Ω μm²), with an approximately 5 ms time constant. However, these numbers began to rise almost instantly (<1 s) and the resistance rapidly became unmeasurably high, with a very long time constant (many seconds) and an unstable baseline.

Electrolytic chloride deposition on to the surface for a few seconds (9 V, with a 100 MΩ current limiting resistance), results in a new, stable DC potential level, ie, a junction potential adjustment of −34 mV, which may simply indicate the removal of the previously unbalanced junction potential between the silver metal and the electrolyte. The new resistance, of the same electrode tip, will now measure about 20 MΩ (varying from 1 to 35 MΩ depending on the extent of the immersion of the tip). This value is fairly stable with a time constant of approximately 5 ms.

Further deposition (eg, 5 min, 9 V through 100 MΩ current-limiting resistance) results in a visible build-up of material on the tip (presumably, mostly chloride), a slightly reduced resistance, and an increase in the time constant.

The chloridation process yields electrodes with a significantly more stable baseline (± 1 mV compared with > 100 mV for non-chloride silver or gold). The measured resistance and capacitance for these electrodes (R_{tip} = 20 MΩ, C_{tip} = 250 pF) are in agreement with our estimates for the tips. These parameters are close to typical values of filled glass capillaries and may permit accurate DC recording.

1.3.3.5   Passivation Layer

To achieve insulation of the needle base we used a thin-film dielectric coating. Because of the topography and fragility of the sharp needles, sputtering deposition is advantageous over spinning of organic material. A convenient way to achieve insulation is by using sputtered silicon nitride. Owing to better durability, silicon nitride is better suited for such applications than silicon dioxide or polyimide [43, 44]. For better corrosion resistance and prolonged device lifetime, additional deposition material, such as triplex layers of silicon nitride and silicon dioxide, will be studied in the future [33]. To verify that our coatings are pinhole free we tested the nitride layers as a mask for aluminium etching. High-quality coatings were achieved for deposition at low background pressure.

1.3.3.6   Experimental Results

To test the performance of our electrodes we first used them in an extracellular preparation of *Manduca sexta* (hawk moth) (see Figure 1.3-11 a). *Manduca* is among the largest of flying insects; its flight control neural circuits are relatively well understood in the context of constrained laboratory environments. This insect has been studied extensively, including its flight dynamics, neuromuscular control, and visual and mechanosensory signaling [45].

A moth was anchored to a holder under a microscope and its lobula plate (the brain optic lobe) was exposed. A conducting wire was connected to the backside of a micro-machined needle device, and the needle was lowered into the lobula plate. A reference electrode was placed at a nearby position. In Figure 1.3-12 a

 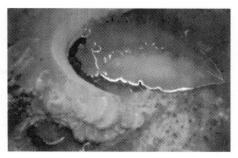

Figure 1.3-11. The biological models used in this work (**a**) *Manduca sexta* (hawk moth) is typically 4 cm in length with a 12 cm wingspan; at 2.5 g, it is among the largest of insect flyers. It can easily carry a test-electronics payload. (**b**) *Tritonia diomedea* (sea slug) is typically 20 cm in length, and has a readily accessible brain with large and well-characterized neurons (see Figure 1.3-1).

we show a record of the evoked extracellular potentials of one neuron in the lobula plate.

To test our gold-coated needles for intracellular applications we used the brain of *Tritonia diomedea* (a sea slug). Unlike the dry setup of the moth experiments, here the isolated brain was anchored in seawater under an optical microscope. An Ag/AgCl reference electrode was dipped in the seawater close to the brain. An insulated conducting wire was connected to the back of the micro-machined needle and the connection was insulated with varnish. At 100 Hz the electrode impedance was of the order of 1 MΩ.

Owing to the enormous dimensions of the sea slug brain cells ($\sim 400\ \mu m$), it is possible to monitor visually the penetration process and select an appropriate location for the probe as it approaches the cell. The micro-machined needles were mounted on a micro-manipulator and were slowly pressed against a cell membrane. Two effects were observed as the needles approached the cell: the measured background potential drifted and the membrane bent. After moderate tapping on the micro-manipulator, spikes were observed. This is probably due to cell membrane penetration. The recorded data are shown in Figure 1.3-13. Similar tests with dull electrodes (tip size $\sim 5\ \mu m$) did not result in signals with impulses.

The noise level seen in Figure 1.3-13 is due to induced 60 Hz interference and possibly to insufficient insulation and grounding of surrounding devices. Biofouling (the affinity of proteins to adhere to synthetic surfaces) may also contribute to the noise and the instability, which was observed in two separate tests of 1 h of recording. Finally, a damaged membrane is likely to contribute to the noise levels and to the relatively small amplitude and slow time constant signal seen in Figure 1.3-13.

The results presented so far show similarity between the geometry of the silicon needles and the pulled glass capillaries (see Figures 1.3-4 and 1.3-9) and support the potential for this process to produce intracellular silicon-based nee-

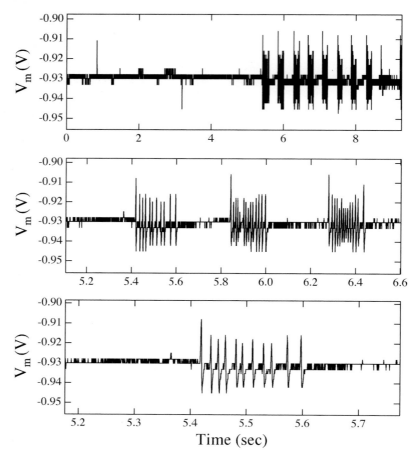

Figure 1.3-12. Evoked extracellular potentials in the lobula plate of *Manduca sexta* (hawk moth) plotted versus time (raw data).

dles. Further, the data in Figures 1.3-12 and 1.3-13 hint at the exciting possibilities for fully fledged neurobiological experiments using silicon-based electrodes.

1.3.3.7   Current and Future Work

The current design of our devices permits convenient handling by using the large base (Figure 1.3-10) to hold and manipulate the needles. During the measurements, however, when the needles are soaked in a conducting medium (eg, seawater or blood), the base acts as a large capacitor (impedance of $\sim 10\,M\Omega$ at 100 Hz). This capacitor is in parallel with the active sensor (the tip of the nee-

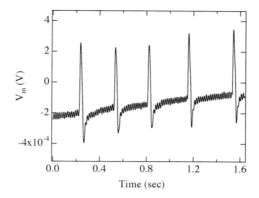

Figure 1.3-13. Recording results with an electrode coated with gold and insulated with nitride. Spontaneous intracellular potentials in a neuron in the brain of *Tritonia diomedea* (sea slug) plotted versus time. The positioning of the probe was controlled via micro-manipulators and an optical microscope.

dle) and can cut off valuable data. A better design should take this into account by limiting the dimensions of the base.

As a first step to improve our process, the dicing saw will be replaced with a deep reactive ion etching (DRIE) process. This will allow a versatile design of needle arrays. Future work will also focus on a versatile connectivity scheme. Preliminary work suggests that polyimide can be used as a convenient structural flexible connecting material. The qualities of the polyimide as a good ionic insulator can be employed to protect metallic lines, which will be used to connect the electrodes and to build large needle arrays. Finally, to enhance the electrode bio-compatibility a non-fouling coating will be deposited. Such a coating was recently tested for bio-MEMS applications and was verified to have good adhesion to silicon, nitride, and gold [9]. Also, it was found that this coating has good ionic conductivity.

1.3.4   Summary

We have reviewed the motivation and the main principles that underlie intracellular potential recording electrodes. We have presented a technique to produce sub-micrometer sharp, high aspect ratio silicon needles. With the refined geometry we were able to obtain high-quality *in vivo* extracellular recordings. Moreover, we presented the first evidence for cell penetration and recording with silicon needles inside a cell. With the advances in bio-MEMS along with the techniques discussed here, the long-term goals of our research are to build stand-alone implantable sensing units made of probes, amplifiers, and memory components, with the specific goal of allowing intracellular recording from freely behaving animals.

1.3.5   Acknowledgments

The authors thank Chris Diorio, Tom Daniel, and Denice Denton for formulating the original concepts that created the basis for this work. They also thank Udo Lang, Jaideep Mavoori, Mark Holl, Buddy Ratner, Vickie Pan, and Jamie Theobald for very useful discussions and Greg Golden, Gary Holman, Joel Reiter, and Xiaorong Xiong for valuable assistance.

This research was supported in part by the David and Lucile Packard Foundation, grant 2000-01763, and a National Sciences and Research Council (Canada) Post-graduate Scholarship to R.C.W. Work in the UW MEMS laboratory by Y.H. and K.F.B. was supported in part by DARPA Bio:Info:Micro grant MD A972-01-1-002, NSF CISE Postdoctoral Research Associateship EIA-0072744 to Y.H. and by Agilent Technologies, Intel Corporation, Microsoft Research, and Tanner Research Inc.

1.3.6   References

[1]   Najafi, K., in: *Proceedings of the IEEE 6th International Symposium on Micro Machine and Human Science*; 1995, pp. 11–20.
[2]   Della Santina, C.C., Kovacs, G.T.A., Lewis, E.R., *J. Neurosci. Methods* **72** (1997) 71–78.
[3]   Bragin, A., Hetke, J., Wilson, C.L., Anderson, D.J., Engel J., Jr., Buzsáki, G., *J. Neurosci. Methods* **98** (2000) 77–82.
[4]   Kandel, E.R., Schwartz, J.H., Jessell, T.M., *Principles of Neural Science*; Norwalk, CT: Appleton & Lange, 1991.
[5]   Alberts, B., Bray, D., Lewis, J., Raff, M., Roberts, K., Watson, J.D., *Molecular Biology of the Cell*; Garland Publishing, New York, NY, 1994.
[6]   Ogden, D., *Microelectrode Techniques, The Plymouth Workshop Handbook;* Cambridge: Company of Biologists, 1994.
[7]   Shepherd, G.M., *Neurobiology*; Oxford: Oxford University Press, 1994.
[8]   Brown, K.T., Flaming, D.G., *Neuroscience* **2** (1977) 813–827.
[9]   Bard, J., Faulkner, L.R., *Electrochemical Methods*; New York: Wiley, 1980.
[10] Robinson, D.A., *Proc. IEEE* **56** (1968) 1065–1071.
[11] Geddes, L.A., *Electrodes and the Measurement of Bioelectric Events*; New York: Wiley, 1972.
[12] Kovacs, G.T.A., in: *Enabling Technologies for Cultured Neural Networks*, Stenger, D.A., McKenna, T.M. (eds.); San Diego: Academic Press, 1994, Chap. 7.
[13] Gesteland, R.C., Howland, B., Lettvin, J.Y., Pitts, W.H., *Proc. IRE* (1959) 1856–1862.
[14] Geddes, L.A., Baker, L.E., *Principles of Applied Biomedical Instrumentation*; New York: Wiley, 1968.
[15] Willows, A.O.D., *Science* **157** (1967) 570.
[16] Sze, S.M., *Semiconductor Sensors*; New York: Wiley, 1994.

[17] Kewley, D.T., Hills, M.D., Borkholder, D.A., Opris, I.E., Maluf, N.I., Storment, C.W., Bower, J.M., Kovacs, G.T.A., *Sens. Actuators A* **58** (1997) 27–35.

[18] Wise, K.D., Angell, J.B., Starr, A., presented at the 8th International Conference on Medical and Biological Engineering, 1969.

[19] Blum, N.A., Carkhuff, B.G., Charles, H.K., Edwards, R.L., Meyer, R., *IEEE Trans. Biomed. Eng.* **38** (1991) 68–74.

[20] Ensell, G., Banks, D.J., Ewins, D.J., Balachandran, W., Richards, P.R., *J. Micro-electromech. Syst.* **5** (1996) 117–121.

[21] Rutten, W.L.C., van- Wier, H.J., Put, J.H.M., *IEEE Trans. Biomed. Eng.* **38** (1991) 192.

[22] Papageorgiou, D., Bledsoe, S.C., Gulari, M., Hetke, J.F., Anderson, D.J., Wise, K.D., presented at the 14th IEEE International Conference on Micro Electro Mechanical Systems, 2001.

[23] Chen, J., Wise, K.D., *IEEE Trans. Biomed. Eng.* **44** (1997) 770.

[24] Kristensen, B.W., Noraberg, J., Thiébaud, P., Koudelka-Hep, M., Zimmer, J., *Brain Res.* **896** (2001) 1–17.

[25] Horbett, T.A., in: *Biomaterials: Interfacial Phenomena and Applications*, Cooper, S.L., Peppas, N.L. (eds.); Washington DC: American Chemical Society, 1982, p. 233.

[26] Costerton, J.W., Stewart, P.S., *Sci. Am.* **285** (2001) 75–81.

[27] Zhang, M., Desai, T., Ferrari, M., *Biomaterials* **19** (1998) 953–960.

[28] Auerbach, H., Soller, B.R., Peura, R.A., in: *Proceedings of the 20th Annual Northeast Bioengineering Conference*; 1994, pp. 108–109.

[29] Yang, Y., Zhang, S.F., Kingston, M.A., Jones, G., Wright, G., Spencer, S.A., *Biosens. Bioelectron.* **15** (2000) 221–227.

[30] Choi, E.T., Callow, A.D., in: *Implantation Biology: the Host Response and Biomedical Devices*, Greco, R.S. (ed.); Boca Raton, FL: CRC Press, 1994, Chap. 3.

[31] Stelzle, M., Wagner, R., Jagermann, W., Fröhlich, R., in: *Proceedings of the 18th Annual International Conference of the IEEE*; 1997, Vol. 1, p. 114.

[32] Hanein, Y., Pan, Y.V., Ratner, B.D., Denton, D.D., Böhringer, K.F., *Sens. Actuators B* (2001). in press

[33] Fassbender, F., Schmitt, G., Schöning, M.J., Luth, H., Buss, G., Schultze, J.W., *Sens. Actuators B* **68** (2000) 128–133; Schmitt, G., Schultze, J.W., Fassbender, F., Buss, G., Lüth, H., Schöning, M.J., *Electrochim. Acta* **44** (1999) 3865–3883.

[34] Regehr, W.G., Pine, J., Rutledge, D.B., *IEEE Trans. Biomed. Eng.* **35** (1998) 1023–1032.

[35] Grattarola, M., Martinoia, S., *IEEE Trans. Biomed. Eng.* **40** (1993) 35–41.

[36] Buitenweg, J.R., Rutten, W.L.C., Marani, E., in: *Proceedings of the 22nd Annual International Conference of the IEEE Engineering in Medicine and Biology Society*; (2000) 2004–2007.

[37] Buzsaki, G., Kandel, A., *J. Neurophysiol.* **79** (1998) 1587–1591.

[38] Campbell, P.K., Jones, K.E., Huber, R.J., Horch, K.W., Norman, R.A., *IEEE Trans. Biomed. Eng.* **38** (1991) 758–768.

[39] Thiébaud, P., Beuret, C., Koudelka-Hep, M., Bove, M., Martinoia, S., Grattarola, M., Jahnsen, H., Rebaudo, R., Balestrino, M., Zimmer, J., Dupont, Y., *Biosens. Bioelectron.* **14** (1999) 61–65.

[40] Chun, K., Hashiguchi, G., Toshiyoshi, H., Fujita, H., Kikuchi, Y., Ishikawa, J., Murakami, Y., Tamiya, E., in: *Technical Digest of IEEE International Conference on MEMS*; (1999) 406–411.

[41] Bengtsson, M., Wallman, L., Drott, J., Laurell, T., in: *Proceedings of the 20th Annual International Conference of the IEEE Engineering in Medicine and Biology Society*; 1998, p. 2229.
[42] Griss, P., Enoksson, P., Stemme, G., in: *Proceedings of the 14th IEEE International Conference on Micro Electro Mechanical Systems*; 2001, p. 46.
[43] Frazier, A.B., O'Brien, D.P., Allen, M.G., in: *Proceedings of the IEEE International Conference on Micro Electro Mechanical Systems;* (1993) 195–200.
[44] James, K.J., Norman, R.A., in: *Proceedings of the 16th Annual International Conference of the IEEE*; 1994, pp. 836–837.
[45] Moreno, C.A., Tu, M.S., Daniel, T.L., *Am. Zool.* **40** (2000) 1138–1139.

List of Symbols and Abbreviations

Symbol	Designation
A	area
A_{Base}	area of passivation layer
A_{tip}	tip area
r	radius
C_d	glass wall capacitance
C_e	electrode capacitance
C_{tip}	tip capacitance
d	thickness of passivation layer
F	Faraday constant
f	frequency
h	height
i	induced current
i_0	exchange current density
l	length
L	length
R	gas constant
R_e	electrode resistance
R_{tip}	tip resistance
T	temperature
V_e	electrode voltage
V_m	resting membrane potential
ΔV	applied potential
α	angle
ε_0	dielectric permittivity of free space
ε_r	relative dielectric permittivity of passivation layer
ρ	resistivity

Abbreviation	Explanation
DRIE	deep reactive ion etching
fMRI	functional magnetic resonance imaging
MEA	microelectrode array
MEMS	microelectromechanical systems
PET	positron emission tomography
RIE	reactive ion etching
ESEM	Environmental scanning electron microscope

1.4 Array Optimization and Preprocessing Techniques for Chemical Sensing Microsystems

D. M. WILSON, S. GARROD, S. HOYT and S. MCKENNOCH,
University of Washington, Seattle, WA, USA
K. S. BOOKSH, Arizona State University, Tempe, AZ, USA

Abstract

Physical sensors, defined by their direct chemical interaction with the sensing environment, are a valuable and often essential contribution to the solution of stringent chemical sensing problems. Methods for conditioning signals from these sensors to optimize their presentation to subsequent decision-making models in the signal processing flow are presented. The assembly of sensors into an array and the preprocessing of these signals are the two primary techniques for conditioning a chemical image for concentration detection, chemical discrimination, and, in some cases, odor localization. Array optimization involves locating the point at which adding additional sensors to an array generates more noise than information and is highly dependent on the application and number of analytes to be sensed or differentiated. Signal preprocessing techniques include noise reduction, feature extraction, the reduction of array inputs, and the scaling of individual sensor signals and provide a means by which to reconstruct an array of chemical sensor signals into a subset of information that enables a decision-making model to do its job with greater efficiency and accuracy. In this chapter, surveys of methods are accompanied by representative examples employing a wide variety of sensors including ChemFETs, chemiresistors, acoustic wave devices, and surface plasmon resonance sensors.

Keywords: chemical sensors; gas sensors; electronic nose; preprocessing; array optimization

Contents

1.4.1   Introduction

Chemical sensing problems are especially challenging for a number of reasons including the complex nature of chemical images as they relate to describing quantity and type of odors and the multidimensional nature of analytes in chemical sensor space. Unlike image processing, a fundamental understanding of the meaning of time and space in output maps of chemical sensor arrays is absent due to the less straightforward nature of multidimensional maps of chemical sensor information. As a result, many successful chemical sensing microsystems that cannot possess full chromatographic or spectrographic analysis capability have relied not on precise understanding of components but on empirical and iterative testing of decision-making models and signal processing techniques. Even when an ideal solution, architecture, and signal processing scheme can be derived from theory, the sensor technology availability can limit the implementation of such a solution. Variations in the ambient environment caused by fluctuations in humidity, temperature, and other factors as well as irreversible reactions between environment and sensor can further hamper the successful implementation of effective chemical sensing architectures.

The practical design of chemical sensing systems involves working within the constraints of available sensor technologies. Arrays of sensors have often been chosen by their ability, through trial and error, to enable a decision-making model to make decisions regarding the sensing environment with sufficient accuracy. However, more rigorous methods for analyzing potential sensor technologies can be generated, analyzed, and evaluated without regard to the choice of a decision-making model. Such optimization techniques, as presented in the technical literature, are reviewed here and representative examples of array optimization methods are presented for arrays of composite film polymer chemiresistors.

Once an array has been constructed, the performance of the sensing system can be further enhanced by preprocessing signals prior to transferring them to the decision-making model for the system. Preprocessing techniques must be chosen to improve the resolution and robustness of the decision-making model and are dependent on the analytes to be analyzed as well as the primary goal of the model. Preprocessing techniques can range from reducing noise within a certain bandwidth to extracting features from raw sensor responses. The preprocessing techniques surveyed here include noise reduction and evaluation, reduction of array inputs, scaling of individual sensor inputs, and feature extraction. Noise in chemical sensing systems can either be eliminated outright using appropriate filtering or aggregation techniques, or it can be quantified and used to validate each sensor in an array. Array inputs can and must be reduced when strong correlation among variables (combination of raw data and extracted features) is compounded by high fluctuations in these variables. Reduction of the inputs using appropriate preprocessing techniques can retain the information in correlated variables while reducing their noise and variability. Scaling of individual sensor inputs, in a similar manner, can reduce noise or superfluous common-mode information in individual sensor signals, including that induced by manufacturing variations, the temperature dependent offsets, and hysteresis, while focusing the attention of the system resolution on the differential mode signals (information used to differentiate one set of signals from another). Appropriate scaling can then enhance system resolution by providing more computational power to the signal range that represents events of interest in the sensing environment. Finally, feature extraction in chemical sensors can involve reducing or supplementing raw data with a set of characteristics that describe sensor response in a manner relevant to the decision-making model at the back end of the signal processing flow. In combination, array optimization and signal preprocessing can provide significant improvement to the performance of decision-making models and is gaining popularity for solving stringent chemical sensor problems using available sensor technologies. In this work, we survey some of the important demonstrations of array optimization and uses of signal preprocessing. To highlight the usefulness of these types of signal conditioning techniques, the discussion of each technique is complemented by a corresponding representative example of successful implementation.

1.4.2   Use of Chemical Sensor Arrays

In an ideal environment, with fixed or no humidity, fixed temperature, and no turbulence of air flow, an array of chemical sensors with overlapping specificity of the order of 8–10 is typically sufficient for discriminating a small number of chemicals or limited mixtures of those chemicals. When the sensing problem or environment becomes more complicated, however, more sensors become necessary to accomplish the same discrimination and concentration detection goals. Additional sensors require adequate array optimization and signal processing techniques to ensure that the signals they provide do not overwhelm the computational capability of the system.

We define two types of arrays for addressing the task of array selection and optimization for a specific sensing application. The homogeneous array consists of sensors that are fabricated alike and operate under the same conditions. For a perfect sensor technology, the outputs of the sensors in a homogeneous array would be identical. For an imperfect but mature sensor technology, the outputs of these same sensors reflect random variations in the sensor surface and sensor fabrication process. For less mature sensor fabrication technologies, the problem of homogeneous array construction is further complicated by systematic mismatch among batches and variations in sensor control parameters. Both systematic and random variations, as long as they are represented by accurate models, can be detected and compensated in arrays of like sensors through calibration and online aggregation of signals. The use of homogeneous arrays to improve the accuracy and resolution of the overlying heterogeneous array is considered a form of sensor preprocessing and is discussed in detail in the next section.

To date, the majority of research effort in constructing chemical sensor arrays has been focused on the construction of the second type of sensor array, the heterogeneous array, for the explicit purpose of enhancing chemical discrimination capability. In the heterogeneous array, all sensors are different whether by nature of their sensor coatings, different operating temperature setpoints, or other variations. Sensors in a heterogeneous array are designed to be different, to separate their specificities in such a way as to improve their discrimination capability. In many efforts directed at the optimization of these heterogeneous arrays, it has become obvious that more information is not necessarily better information. Often, array optimization techniques determine that a moderate number of sensors is best, as noise and information reach a trade-off at a point of peak performance. For the same reasons, more information from each sensor is not necessarily better information. For heterogeneous arrays, then, it is as important to potential communication bottlenecks downstream of the sensors as it is to the chemical discrimination and concentration detection problems to optimize both number of sensors and the information extracted from each sensor.

Attempts to model the precise specificity of chemical sensors assembled into arrays have been thwarted by the sheer complexity of the number of interactions between the ambient environment and the sensor surface. As a result, many research efforts have relied on qualitative or empirical design of the array, based on the success of the decision-making model in processing the array outputs. Typically, arrays are assembled using sensors that, individually, have known preferences (selectivities) for one or two analytes of interest, so that, in combination, the arrays demonstrate selectivity to the mixture or set of analytes of interest. Low preference or selectivity for anticipated interferents in the sensing application also plays an important role in selecting sensor technologies for the array. Construction of heterogeneous arrays in this qualitative manner has proven to be successful in a variety of applications in measuring both sets of known, single analytes and for measuring complex mixtures for which all the component analytes may not be known.

Gardner et al. at the University of Warwick have demonstrated the use of chemical sensor arrays for single analyte and complex mixture recognition in published research efforts beginning in the late 1980s and early 1990s. Efforts to provide sufficient variability in sensor technology and operation for proper array optimization include the modeling of reactions in the most popular chemiresistive sensor, the tin oxide sensor [1], manipulation of such design parameters as electrode geometry (in phthalocyanine [2] and metal oxide thin films [3]), use of multiple modes of operation for each sensor to minimize the physical size of a heterogeneous array architecture [4], the use of molecular sieves at the front end of the sensor array to provide prefiltering (by size) of molecules of interest [5], and the development of improved supporting infrastructure in the form of improved headspace samplers [6]. Various (heterogeneous) combinations of metal oxide and conducting polymer sensors have been used to differentiate single analytes such as toluene, n-propanol, n-octane, methanol, ethanol, 2-propanol and 1-butanol [8]. Heterogeneous arrays of metal oxide and conducting polymer sensors, arranged in different combinations, have also been applied to the study of complex mixtures that are not necessarily fully characterized or understood.

Examples include the discrimination of (1) coffees [9], (2) tainted water [10], (3) alcohols, tobacco blends, and beverages [11], (4) banana ripeness [12], (5) types of beer [13, 14], and (6) paper quality [15]. In combination with the University of Neuchâtel, the Warwick group has also made strides toward reducing power in these arrays to 30 mW per sensor and enhancing the ability to maintain stable operating temperature, both important steps toward microarray implementations of these same arrays [16]. The Warwick group has also carried out fairly extensive studies of pattern recognition techniques optimized to detect the chemicals or mixtures of interest in a given array architecture. In these efforts, the need to optimize the array architecture at the beginning of the design process have not been particularly evident, since the choice and number of sensors in each array have been sufficient to demonstrate the feasibility of using the array for a wide variety of applications. In the broad and overlapping specificities that characterize metal oxide and conducting polymer sensors, the discrimination capability of the array has exceeded the needs of the application. Long-term studies of stability, humidity, and the impact of variable ambient environments, however, have not been extensively addressed in this series of research efforts, since these sensor arrays have often been developed with the assumption of a controlled sampling space at the front end to minimize variations in the sensing environment. The choice of which types of sensors to use in each array have largely been focused on using at least one sensor to represent or sense every analyte of interest in the array. Multiple sensors for the same analyte have been chosen when interference from related analytes or from ambient variations such as humidity are anticipated in the application of the array.

Researchers in the gas sensing group at NIST (National Institute of Standards and Technology) have focused their efforts toward the development and optimization of arrays of metal oxide (and similar chemiresistive materials) sensors for microfabricated systems. Although materials are not necessarily optimized in the formation of sensors in the array, operating temperatures and pulsed sequences of operating temperatures are optimized to distinguish among closely related analytes [17, 18]. Operation in such a 'temperature-programmed' mode permits the tailoring of reaction kinetics in time and space to particular applications. This array design methodology has been applied to the detection of hazardous agricultural chemicals [19], and is suited to a wide variety of gas detection applications involving reducing gases.

Efforts targeted specifically at array optimization using chemiresistive sensors based on a wide variety of conducting polymers and their composites have been conducted at the California Institute of Technology by Lewis et al. In this research, various combinations of the conductive agent, carbon black, and insulating polymers are assembled into conductive chemiresistor arrays in large numbers to obtain both wide discrimination capability and wide dynamic range (number of components to which the array is sensitive) [20–22]. Although not yet fully explored, a broad choice of sensors, linear sensor characteristics, and additive response characteristics of multi-analyte mixtures provide an excellent foundation for the rigorous design of heterogeneous arrays for particular applications in this research effort. Optimization of the arrays to include combinations of homogeneous and heterogeneous sub-arrays on the same, finite-size silicon substrate are anticipated to find the optimal trade-off point between array discrimination capability and low-noise characteristics.

Efforts directed toward assembling arrays of surface acoustic wave (SAW) devices for sensing of chemical warfare agents, their simulants and anticipated interferents have been extensively addressed at Sandia National Laboratories. These efforts have been directed toward developing a portable instrument for field application to the detection of chemical warfare agents. Because of the high power and electronics overhead involved in transducing acoustic wave propagation velocity to an electrical signal as well as general portability concerns, it has been essential to reduce the size of the array as much as possible. From a possible array containing dozens of coatings on top of bare quartz SAW substrates (including the use of bare quartz as a reference coating), these arrays have been rigorously optimized to chemical warfare agent detection. All possible combinations of these dozens of coatings have been evaluated for discrimination accuracy, both in 'new' sensor conditions and in 'old' (reduced sensitivity) sensor conditions [23, 24]. Optimization is performed using a unique form of nonlinear pattern recognition (called VERI) [25] based on the ability of the human eye to separate clusters of information in multidimensional space. Array optimization has led to the selection of seven heterogeneous coatings in the array (bare quartz, two polymer, two metal, one dendrimer, one metal oxide) that are in the process of being implemented into a series of portable systems for the targeted application (detection of concentration and discrimination among interferents of chemical warfare agents). Similar efforts at Sandia are under way to optimize integrated arrays of SAW devices for laboratory-on-a-chip applications on gallium arsenide substrates [26–28]. The optimization of arrays in these efforts via evaluation of all possible combinations of sensors is a significant step toward rigorous array architecture design for chemical sensing applications. In the Sandia efforts, sensitivity loss is modeled as a random loss in sensitivity of all sensors in the array. Evaluation and modeling of typical sensitivity loss over the lifetime of these sensors has not yet been completed but will make an important addition to the array optimization effort. In addition, because of the size and complexity of the support electronics, the addition of homogeneous or redundant sensors to the array is not often practical. The need for signal conditioning and impedance matching in these arrays, however, is recognized and is part of the ongoing, larger research effort.

Efforts at the University of Washington have addressed the benefit of using homogeneous arrays for redundancy and for noise reduction in the signal preprocessing flow associated with arrays of chemical sensor arrays. Compact circuits for removing outliers from homogeneous arrays have been demonstrated [29] for cases where sensor failure is extreme or catastrophic, resulting in an outlying output signal in a relatively short time period. The tendency of variance of discrete tin oxide sensors to decrease proportional to the square root of the number of sensors in the array has also been demonstrated in these efforts [30]. Both removal of outliers and averaging of sensor signals are first steps toward effectively processing homogeneous arrays of sensors into single aggregate output signals (one per homogeneous array) that are more robust and stable than those produced by individual sensors. Efforts to optimize the size of homogeneous arrays within a larger heterogeneous array of metal oxide and polymer sensors, given a fixed overall array size, are ongoing at the University of Washington.

1.4.3 Chemical Sensor Array Optimization

Despite the vast number of research efforts published in the technical literature, demonstrating the effective use of arrays in solving chemical sensing problems, quantitative array design is rare in chemical sensor array research and development. The explicit and quantitative compression of a larger set of available sensor technology types into an optimized, smaller array for application to a particular sensing problem is difficult because of the complexity of chemical sensors and the many influences that affect their response in variable sensing environments. In many cases, the number of different sensor types is limited and does not permit expansive optimization; in other cases, appreciable selectivity is acquired with available technology types and optimization is not necessarily required for demonstration of array performance. Some efforts, such as those by Lewis et al. [21, 22, 31], suggest that larger arrays are always better for maximum chemical diversity desired of many multipurpose "electronic nose" applications. A few groups have explicitly addressed the array optimization problem in a quantitative manner. Gardner and Barklett [32] define performance metrics that includes a resolution power metric for array optimization based on the work of Neibling and Muller [33], for the standardization of diverse chemical sensing systems or "electronic noses"; standardization is notably absent form this research community and is absolutely essential to uniting results produced from the development of sensor technology types, arrays, decision-making models, and application-based implementations for the sake of performance comparison. Chaurdy et al. [34] define another method for selecting an optimum sensor array based on principal components analysis and directed random search algorithms. These methods successfully identify and remove sensors in an array of 80 sensors; the eliminated sensors are those that produce the most significant improvement in sensitivity and selectivity of the array when they are removed from the decision-making model.

As a representative example of array optimization, an array of 10 possible composite carbon black-polymer films is optimized by maximizing the resolution factor between a specific pair of analytes. The polymer films have been fabricated and tested in the Lewis laboratory at Caltech and consist of composite polymer films deposited on gold-coated glass slides. The composites are prepared by combining a conductor (carbon black) with the chemically sensitive (but insulating) polymer in solvent and depositing the mixture on slides via a dipping or airbrush technique. The choices for construction of these films into sensor arrays are 10 different (heterogeneous) sensor films as follows:

1 PMODS (poly(methyl-octadecyl siloxane)
2 Polychloroprene
3 Polyvinylpyrrolidone
4 Polyepichlorohydrin
5 Polycaprolactone
6 DEGA (diethylene glycol adipate)
7 PEVA (18% VA) (polyethylene-co-vinyl acetate (18% vinyl acetate))
8 Polystyrene-butadiene
9 PEVA (45% VA)
10 Polyethylene oxide

Resistance of each sensor is measured as it responds to a vapor introduced in a flow chamber in a background of compressed air. These raw resistance outputs are first converted to signals R_i that represent the change from baseline value during the relevant experiment window and then normalized with respect to all sensor responses at a given time in a given array:

$$R_i = \frac{\Delta \text{ Resistance}_i}{\sum\limits_{i=1}^{M} \Delta \text{ Resistance}_i} \qquad (1)$$

for each sensor i in the array. Each experiment begins with a 300 s period of air flow to purge the sensor surface, a 240 s period during which a solvent vapor is introduced, and a final 300 s purging period to enable each sensor to return to its baseline state. Sensors are tested, in 25 separate experiments, on five solvent vapors: acetone, ethanol, hexane, tetrahydrofuran (THF), and toluene. Resistance at the chosen concentration ranges is linear as a function of concentration and the effect of multiple vapors is additive. Further details of the fabrication and characteristic behavior of these sensors can be found in [20–22]. Primary analytes and interferents are selected from the five possible solvent vapors (acetone, ethanol, hexane, THF, and toluene) in any combination. Array optimization is performed on the principle that the variance ('noise') of a cluster in multidimensional chemical sensor space is directly related to array size for sensors that provide additional useful information to the selective capability of the array. The point at which the additional noise provided by adding a sensor is no longer offset or exceeded by the additional information provided by the array is the optimization point. The metric that defines this trade-off point is called the resolving power of the array and is similar to a metric presented by Gardner and Bartlett [32] and Niebling and Muller [33] in their efforts to standardize metrics for evaluating the performance of an "electronic nose". The resolving power (RP) is defined as follows:

$$RP = \frac{\mu_{d(i,j)}}{\sigma_{d(i,j)}^2} \qquad (2)$$

where $\mu_d(i,j)$ is the average distance from the mean of the first cluster (i) to all points associated with the second cluster (j). In this example, the first cluster is associated with the primary analyte of interest and the second cluster is associated with an interferent with that primary analyte. The distance between the first cluster mean and second cluster members is a Euclidean distance in multidimensional space as follows:

$$\mu_{d(i,j)} = \frac{\sum\limits_{j=1}^{N} d_{ij}}{N} = \frac{\sum\limits_{j=1}^{N}\left(\sqrt{\sum\limits_{k=1}^{M}(\mu_{ik} - R_{jk})^2}\right)}{N} \qquad (3)$$

where N is the number of experiments used to optimize the array (25 in this case), M is the number of heterogeneous sensors (a maximum of 10), R_{jk} is the response of sensor k during experiment j for the second cluster under consideration, and μ_{ik} is the mean of the sensor k response for the first cluster under consideration. The variance parameter $\sigma_{d(i,j)}{}^2$ is the variance in these distances between the first and second clusters:

$$\sigma_{d(i,j)}{}^2 = \sum_{j=1}^{N} \langle \mu_{d(i,j)} - d_{ij} \rangle^2 \qquad (4)$$

for the cluster i under consideration. For 10 possible types of conducting polymer chemically sensitive films, 10! or 3 628 800 possible combinations of sensors are possible for a 10-element array of these sensors. Analysis of this size of array possibilities is computationally prohibitive; as a result, a set of randomly select 100 permutations of these 10 sensors (without replacement) are chosen for analysis. In this example, the array is optimized for detecting one analyte (THF) in the presence of one interferent (toluene); however, the results can be extrapolated to any number of possible analytes of interest and potential interferents.

To optimize the array, the distance between clusters ($\mu_{d(i,j)}$), variation in that distance ($\sigma_{d(i,j)}{}^2$), and resulting performance metric *RP* need to be considered for only one pair of data sets. The maximum *RP* determines the optimal array composition.

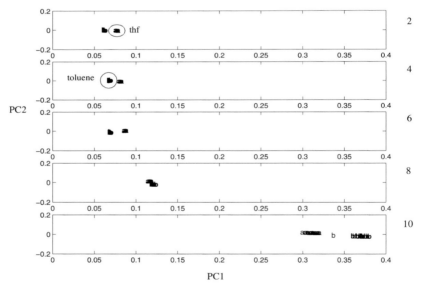

Figure 1.4-1. Array optimization example. Twenty-five experiments for the primary analyte, **(a)** THF, and the interferent, **(b)** toluene, are shown in principal component space as a function of varying array size for the sensor sequence: 1, 9, 8, 10, 7, 3, 6, 5, 2, 4. The optimal array size is six (consisting of sensors 1, 9, 8, 10, 7, and 3). Below this optimal array size, the distance between sensors is smaller, and above this optimal array size, the variance within each sensor cluster increases drastically. The point at which the ratio of these two parameters (mean distance between clusters/mean variance in distance between clusters) is at a maximum is the optimal array composition.

One hundred random permutations of array compositions (without repetition) are considered for each analyte-interferent pair and the array composition that results in the maximum *RP* is the optimal array structure. The results of array optimization for detecting THF in the presence of the interferent toluene indicate that the maximum *RP* occurs for an array size of six sensors containing sensors of types 1, 9, 8, 10, 7 and 3. An array size with less than six sensors results in a decreased distance between analyte clusters, while array sizes above six sensors result in an increased variance in the distance between clusters ($\sigma_{d(i,j)}{}^2$) without a comparable beneficial increase in intercluster distance ($\mu_{d(i,j)}$). This emergence of the optimal array size for a given sequence of sensor types presented to a potential sensor array is shown graphically in principal component space in Figure 1.4-1.

1.4.4   Preprocessing of Chemical Sensor Signals

Preprocessing for sensor systems can be loosely defined as the process of preparing sensor signals for transfer into a decision-making model in such a way that the accuracy of the model is enhanced by the preprocessing. The choice and design of preprocessing techniques are dependent on the type of task to be accomplished (concentration detection or chemical discrimination), the constraints of the implementation mode of preprocessing and decision-making model (eg, hardware or software, serial or parallel information processing), and the type of decision model (eg, cluster analysis, artificial neural network). Most preprocessing techniques can be placed into one of the following four categories:

- Noise evaluation and reduction: elimination or separation of information that is not directly useful for the decision-making model in the chemical sensing system at hand. Noise that is of no use to subsequent signal processing can be removed with the appropriate filtering technique or lock-in amplification (to focus on a frequency of interest). In some cases, noise may be useful to the subsequent signal processing for decision making and it can be evaluated rather than reduced or eliminated to improve the accuracy of decision making at the back end of the signal processing system.

- Reduction of array inputs: in time or space to improve the computational efficiency of the decision-making model. In an array of sensor signals, spatial or temporal dimensionality can be reduced through various filtering or axis transformation techniques.

- Scaling of individual sensor inputs: to focus the full resolution of the subsequent signal processing on the chemical sensing problem at hand. Scaling can eliminate temperature effects, manufacturing variations, and other variables that are not directly relevant to chemical sensor response. Autocalibration of sensors involves scaling of individual sensor outputs to compensate for drift and aging, both linear and non-linear.

- Feature extraction: of salient features of sensor and array responses to the chemical sensing problem at hand. Feature extraction techniques may or may not reduce the input space; instead, feature extraction enables raw data to be transformed into a set of features that are more distinguishable among the classes set forth by the decision-making model for the sensing system.

A combination of techniques from these four categories can be used to improve the stability and accuracy of sensor signals submitted to a decision-making model for chemical concentration detection, discrimination, or, in rare cases, localization. Most research efforts directed at solving chemical sensing problems have used a form of preprocessing to improve the accuracy of the system output.

1.4.4.1   Noise Evaluation and Reduction

In electronic systems, noise reduction typically involves the removal of various random fluctuations in signal output from the net electronic signal output. Electronic noise arises from a variety of sources and often limits the ultimate resolution of an electronic circuit. The most common types of electronic noise are as follows:

- *Thermal noise* is caused by the random movement of electrons and other current carriers in response to thermal excitation. While all materials operating above 0 K exhibit some form of thermal noise, this form of noise is especially troublesome in resistors. By increasing the temperature of operation, power consumption in resistors can increase electron thermal velocity and the associated thermal noise even further than expected for the mean operating temperature for a circuit. Thermal noise is independent of frequency of operation [35] and a typical 1 kW resistor at room temperature operating in a circuit with a 10 MHz bandwidth can generate r.m.s. voltage noise of the order of 10 μV.

- *Shot noise* can be best understood by understanding the fundamentally discrete nature of current flow. At a macroscopic level, current is viewed as a continuous rate of flow of carriers. However, at the atomic and sub-atomic levels, the rate of charge is more discrete because a single electron has only a finite amount of charge. This discrepancy between a discrete and a continuous flow is especially noticeable at low frequencies. Discontinuities in electron movements male shot noise especially prevalent in diodes and bipolar junction transistors. Shot noise is also present in field effect transistors, but is only associated with the gate current and is considered very small [35]. A typical diode operating in a circuit of bandwidth 10 MHz and current of 1 mA can exhibit r.m.s. current noise of the order of 1 nA.

- *Flicker noise* is caused by the interaction of electrons with interface and surface states. In these incomplete states, electrons can become trapped for a longer period of time than in the bulk of the material, producing an additional noise source at frequencies typically below a few kHz, although this boundary can be much higher. Flicker noise is also known as $1/f$ noise, because it exhibits an inverse relationship with frequency. It is a particular problem in field effect transistor (FET) structures, as electrons are accelerated through a high electric field through the channel of the transistor, causing electron traps in the interface between the channel and insulating layer to have a significant effect on the net flow of electrons through this shallow channel underneath the gate. Although it is difficult to provide typical values of flicker noise due to the substantial impact of random crystal imperfections on device constants, a FET characterized by a DC drain current of 1 mA, operating frequency of 1 kHz, and circuit bandwidth of 1 MHz might experience a typical r.m.s. drain current noise of the order of 1 μA [35].

As is the case with many types of sensors, chemical sensors can exhibit fluctuations or noise due both to electronic noise that depends on the mode of transduction from chemical to electrical signal and to interaction with the sensing environment. Unlike purely electronic systems where noise is seldom desired, noise in these chemical sensing systems can be either reduced or used as an evaluation tool to aid in subsequent signal processing. For example, thermal noise fluctuations in combination with environmental fluctuations in a chemiresistor can be evaluated at baseline (no stimulus present in the sensing environment) to verify that the chemiresistor continues to operate primarily as a resistor in a mode similar to its original calibration state. Deviations from the calibration state can be caused by significant sensor drift (which limits the dynamic range of the sensor and the usefulness of the calibration state in making decisions based on the sensor output) or by the introduction of an alternative current path due to poisoning of the sensor by persistent or unexpected chemical influences in the sensing environment. In either case, deviations in the distribution of chemiresistor output signal from the distribution at baseline acquired during calibration can be used to decrease the contribution of the deviant sensor to a system-level sensing decision or can indicate the need to replace the sensor and recalibrate the system to its replacement.

A comparison of healthy sensor behavior with unhealthy sensor behavior for a chemiresistor is shown in Figure 1.4-2. The composite carbon black-DEGA polymer film exhibits a Gaussian distribution of signal during the calibration state (Figure 1.4-2a). Sometime later during sensor operation, this sensor continues to exhibit a Gaussian distribution within a 95% confidence interval (Figure 1.4-2c) according to the single sample Kolmogorov-Smirnov goodness-of-fit hypothesis test. However, at another time, when the sensor is temporarily poisoned by its exposure to ethanol during a recovery period (Figure 1.4-2b), the distribution of sensor signal is not sufficiently Gaussian to fall within the requisite 95% confidence interval that is representative of the baseline condition. At this time, the sensor is flagged as potentially unhealthy and the confidence of the resulting system-level decision (discriminating ethanol from related chemicals acetone, THF,

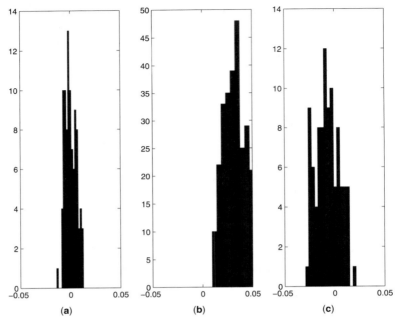

Figure 1.4-2. Preprocssing of chemical sensor information using noise evaluation techniques (histograms). In this example, random distributions of the output of a chemiresistor are evaluated to determine the resulting health of the sensor signal. During calibration (**a**) the sensor histogram exhibits a Gaussian distribution of output signal (resistance) that is expected of the chemiresistor. During a recovery period (**b**) (after exposure to the chemical ethanol), the sensor exhibits a distribution that is sufficiently outside the calibration distributions to raise a warning flag, indicating that its contribution to the decision making model should be considered with less confidence. If the sensor continues to demonstrate a noise distribution significantly outside of its calibration distribution (in the presence of no chemical stimulus), it should be recalibrated or replaced. Finally, during normal sensor operation after sufficient recovery time from exposure to the analyte ethanol, (**c**) the sensor output distribution is restored to a Gaussian distribution at its output within a 95% confidence interval. Each *x*-axis represents the percentage resistance change from the average baseline condition. Each *y*-axis represents the quantity of points at each value (histogram).

and toluene) is reduced. In this case, separation of random noise generated by thermal effects and arbitrary interactions with the sensing environment from the stable (DC) portion of the transduced sensor signal provides a means to determine the health of the sensor. Sensor health can then be used to adjust the level of contribution the sensor makes to system-level operation.

Noise has also been used to supplement information provided for discriminating differences among various compound analytes. Kish et al. [36] use the frequency spectrum of the noise characteristics in an array of Taguchi metal oxide sensors to discriminate differences among tea leaves, potato chips, and white pepper. Within the frequency spectrum, the mean value of the slope in each frequency decade is used as the salient information with which these three analyte mixtures are differentiated.

As is the case in many electronic systems, it is often desirable to reduce or eliminate noise altogether in chemical sensors rather than retain it for evaluation. Noise reduction is frequently used in situations where multiple sources of the noise cannot be sufficiently separated to use this part of the sensor signal as an evaluation tool. In these cases, noise is no longer useful to the functionality of the system and can be minimized using averaging or filtering techniques. An example of the usefulness of averaging (spatial filtering) is shown in Figure 1.4-3 for a homogeneous array of tin oxide sensors. Between 2 and 40 Taguchi tin oxide sensors of the same type and fabrication are averaged to obtain an aggregate output whose variance, as expected for a Gaussian distribution, decreases as the inverse of the square root of the number of sensors in the array:

$$\sigma_{array}^2 \propto \frac{1}{\sqrt{N}} \tag{5}$$

This reduction in variance, accomplished through the redundancy of a homogeneous array, has a direct effect on the resolution of the chemical sensing system. The nature of this effect, whether it be linear or nonlinear, will depend on the de-

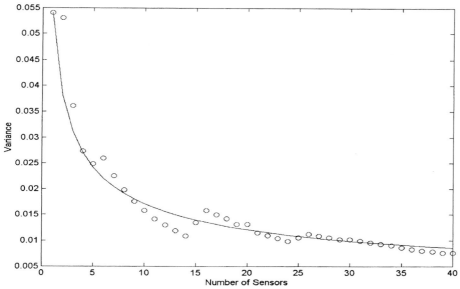

Figure 1.4-3. Preprocessing using noise reduction techniques. The effect of averaging like sensors in a homogeneous set (same material, same fabrication method) is shown for powder-based tin oxide sensors. As expected for a Gaussian distribution, the variance of the aggregate output (average of between two and 40 sensors) decreases as the square root of the number of sensors in the array.

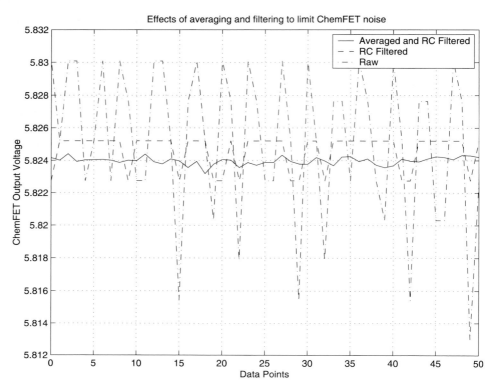

Figure 1.4-4. Preprocessing of chemical sensor information using noise reduction techniques. The voltage output of a ChemFET coated with a chemically sensitive polymer is shown before preprocessing, after filtering, and after filtering and time window averaging. A constant current flows through the FET, as controlled by external circuitry. The drain and gate of the ChemFET are connected together, resulting in an output voltage dependent on the FET characteristics and also work function changes induced by the presence of certain chemicals in the external sensing environment. Before filtering, significant noise is present, with a standard deviation of 4.8 mV. After low-pass filtering, the standard deviation of the noise is reduced to 1.20 mV. Finally, after both low-pass filtering and time-window averaging, the noise varies by a standard deviation of 0.32 mV.

cision-making model used to determine concentration, discriminate chemicals, or localize odor sources in the system.

Further reduction of noise in chemical sensor signals can be accomplished by temporal filtering. Temporal filtering smoothes a signal over time so that when it is digitally sampled, fluctuations of that signal are reduced. A polymer-coated ChemFET is used to demonstrate the usefulness of temporal filtering. The ChemFET is a standard *n*-channel FET constructed in silicon, diode-connected (drain is connected electrically to the gate of the transistor), and coated with a customized polymer matrix during post-processing. The source of this noise is suspected to be caused in part by device physics (specifically thermal and flicker noise due to the large transistor, large operating drain currents, and low operating frequencies) and due to data acquisition methods. Noise generated by data acquisition methods is caused by the limited resolution of analog-to-digital conversion, the type of analog-to-digital conversion (successive approximation), limited sampling rate (remaining noise may be aliased), and radiofrequency interference inducing noise in unshielded wiring. Experiments have shown that noise reduction is required for these sensor signals to permit adequate resolution and signal-to-noise ratio. It is not possible, within the constraints of the fabrication process, to efficiently examine, isolate, and separate the individual sources of this noise. Since the signal is read at DC, however, these sources of noise can be easily filtered out using a first-order, low-pass filter with a cut-off frequency of 8 Hz. Additional smoothing is then done within the data acquisition software. The resulting signal (Figure 1.4-4) demonstrates significantly reduced noise, thereby enhancing the signal-to-noise ratio and system resolution.

1.4.4.2   Reduction of Array Input Space

Reducing the array input space, if done without significant loss of information, has obvious implications for reducing the amount of computation performed by the decision-making model at the back end of a chemical sensing microsystem. In some cases, fewer computations can enable the use of pattern recognition or decision-making models that could not ordinarily converge or operate with a larger number of inputs. The size of the input space is significantly larger than the number of inputs itself and can increase exponentially with the number of inputs [37]. The input space is the possible number of patterns that can be generated from a certain number of inputs. As the input space increases, the amount of data required for the decision-making model to train adequately for generalizing to vectors that it has not seen before also increases substantially. The development (training) of a valid decision-making model becomes increasingly difficult with an increasing input space, due to both the increase in amount of training time and the amount of data required to complete a valid training cycle. As the size of the input space increases exponentially with the number of inputs, it is fairly easy to reach the point in chemical sensing array space, because of array size and problem complexity, where training time is prohibitive under the constraints of modern microcomputer clock rates.

The size of the input space also impacts the operation of the model after training. With many nonlinear pattern recognition models, such as the wide variety of available artificial neural networks (ANNs), the number of processing nodes increases faster than the number of inputs. For example, a standard three-layer, perceptron-style artificial neural network with N (seven) inputs, $2N$ (14) hidden layers, and two outputs requires approximately 142 computations (assuming multiply/add is one computation and including the activation functions at hidden and output neurons); decreasing the inputs N to six results in an ANN that requires only 110 computations, a 22.5% reduction compared with the 14% reduction in inputs from seven to six. In almost all cases, justified reduction of the input space, by reducing the number of computations required to make a decision, improves the speed, reduces the power dissipation and increases the portability for a sensing instrument or system. In the development of a portable sensing instrument, reduction of the input space not only improves the quality and feasibility of training a decision-making model to process adequately inputs from the system, but also improves the overall performance characteristics of the instrument (battery life, size, and real-time operation).

Reducing the input space in a chemical sensing system can also reduce the amount of noise, or superfluous information presented to the decision-making model, thereby increasing its accuracy. Variables or sensors that provide similar information as provided by other sensors in the array can be eliminated in systems used for chemical discrimination. Elimination of highly correlated variables, when the correlation exceeds the additional information provided by multiple variables, effectively reduces the noise in the array without sacrificing accuracy or performance.

Finally, reduction of the input space can improve the visualization of the data, temporarily for evaluation purposes or permanently for decision-making purposes. Since the meaning of neighboring relationships among sensors in a chemical sensor array is often not straightforward, the selection of an appropriate decisionmaking model often relies on human visualization of the data in multi-dimensional space. Programs are available to view dimensions greater than three; however, projections of data on to two or three dimensional principal component space are the most common method for visualizing chemical sensing data to demonstrate the usefulness of an array, to evaluate the improvement provided by preprocessing techniques, or to determine an appropriate decision-making model for the data. Analyses of these principal components and the clusters that are formed by the primary principal components have been used on a widespread basis to make discrimination decisions regarding a wide variety of analytes.

An example of the benefit provided by reduction of the input space in a chemical sensing system is demonstrated for an array of seven SAW devices applied to the discrimination of 13 different, vapor-phase analytes. Each SAW device consists of a series of interdigitated electrodes on quartz coated with a chemically sensitive material (bare quartz, dendrimer, polymers (SE-30, PECH), and metals (CuO, Au, Pt)). The SAW device is connected in an oscillator loop using an operational amplifier and associated circuitry and allowed to resonate. The resonant frequency is measured as an output indicative of analyte concentration and changes with the speed of the propagating acoustic wave between the device electrodes. The acoustic wave speed and resulting shift in resonant frequency of the measurement circuit vary with the SAW device coating and with the type and concentration of analyte in the sensing environment. The array of seven surface acoustic wave devices is exposed to 13 analytes in vapor phase in a controlled testing chamber in concentrations varying from 0.40 to 48.0 P/P_{sat}, where P and P_{sat} are the partial pressure and saturated partial pressure of the analyte of interest, respectively. These analytes are a combination of aliphatic, aromatic, and chlorinated hydrocarbons, alcohols, ketones, and organophosphorus compounds:

(a)　kerosene;
(b)　benzene;
(c)　toluene;
(d)　chlorobenzene;
(e)　carbon tetrachloride;
(f)　trichloroethylene (TCE);
(g)　methanol;
(h)　propanol;
(j)　pinacolyl alcohol;
(k)　acetone;
(m)　methyl isobutyl ketone;
(p)　diisopropyl methylphosphonate (DIMP);
(r)　dimethyl methylphosphonate (DMMP)

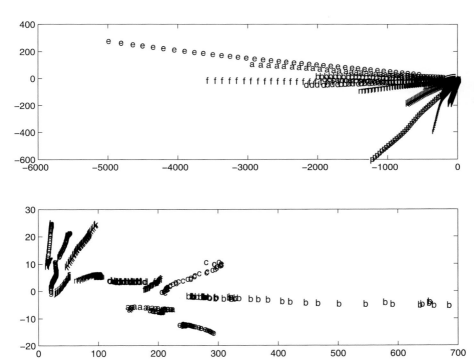

Figure 1.4-5. Preprocessing using the reduction of array input space. Shown are the results of projecting the first two principal components (first principal component on the *x*-axis; second principal component on the *y*-axis) of the SAW device data into two-dimensional space using (**a**) seven data dimensions and (**b**) six data dimensions. Reduction in the input data (by dividing all sensors by the bare quartz SAW reference output) provides less data and more differentiation capability to the decision-making model at the back end of the system. The x and y axes are the first and second principal components of the array data respectively.

The objective for the decision-making model in this system is to discriminate among the 13 analytes tested. Reduction in the input space can be accomplished by dividing all the outputs of the system by the output for the bare quartz SAW device at every point in time. The bare quartz coating exhibits no significant differences in response among analytes and is, therefore, a good choice to serve as reference coating for reducing the input dimensionality from seven to six. Before reduction (Figure 1.4-5 a), analyte data, when projected on to the first two principal components in two-dimensional space, are difficult to differentiate, particularly at low concentrations. When reduced to six dimensions by taking the ratio of all data to the reference (Figure 1.4-5 b), however, the data become significantly easier to differentiate, in large part owing to the reduction of the influence of concentration in the discrimination process. The drift that remains in the data after this reduction step is due to nonlinearities and differences among analytes in the response (calibration curves) of the array to each analyte across the full range of concentrations.

1.4.4.3   Scaling of Individual Sensor Inputs

By the very nature of their heterogeneity, arrays of chemical sensors can exhibit a wide range of baseline values, types of outputs (resistance, voltage, etc.), and dynamic ranges. An important role of electronic preprocessing is to reduce the range of the inputs extracted from these arrays. Such reduction generates a more uniform data set so that the capabilities of the electronic signal acquisition are focused on the actual signal range rather than on artifacts of the sensor transduction. In this section, we discuss two important techniques for focusing the strength of the electronic signal acquisition on signal rather than artifact: baseline compensation and response normalization.

Many chemical sensors that physically interact with the environment are subject to wide fluctuations in baseline output partly because of the difficulty involved in controlling surface characteristics during the fabrication. Surface characteristics are notoriously difficult to control during microfabrication and, in most electronic devices, the signal due to surface characteristics is minimized. In chemical sensors, however, it is the surface that provides most if not all of the sensor signal in response to chemical changes in the sensing environment. In order to acquire a useful response of the sensor, the surface must be a significant part of the overall electronic signal in comparison with the bulk. This paradox of chemical sensor fabrication results in large variations in baseline characteristics for good chemical sensors. In many chemical sensing systems, variations in baseline are not compensated and subsequently consume a large part of the resolution of the signal processing electronics. The straightforward, brute-force solution to this problem is to increase the resolution of the signal processing electronics to a number of bits that accommodates the range of baselines generated by fabrication variations and the required resolution within the sensor response itself. A more elegant solution is to compensate for these baseline variations during signal preprocessing so that subsequent analog-to-digital converters in the signal flow focus their resolution capability on the sensor response itself. An example of such baseline compensation is shown in Figure 1.4-6. Before compensation (Figure 1.4-6a), the baseline resistance of the poly(ethylene oxide) chemiresistor varies from 25 to 300 kΩ for different sensors in the same fabrication batch. The sensor resistance is extracted as a voltage by injecting a constant current into the resistor and measuring the resulting voltage across the chemiresistor. To demonstrate the effectiveness of baseline compensation, consider a sensor array that uses a constant current of 50 mA and exhibits baseline resistance changes of as much as 20%. The range of output voltages, under normal operating conditions and these baseline characteristics, is 1.25–18 V. When accompanied by a 12-bit analog-to-digital converter (ADC) at the next stage of signal processing, the minimum detectable change in either sensor is 81.8 Ω (0.327–0.027%). When using compensation (Figure 1.4-6b), the minimum detectable change is 2.4–29.3 Ω (0.010%), produced from a voltage range of 2–2.4 V. As another way of evaluating the benefits of baseline compensation, an ADC of more than 17 bits is required to achieve the same resolution in an uncompensated array, compared with 12 bits in a compensated array. Compensation provides up to a 30-fold improvement in response (eg, concentration) resolution for the same sensors and electronic processing.

Another common technique for scaling individual sensor inputs is the normalization of sensor input data with respect to other sensors at the same time or with respect to the same sensor across all response time.

An example of the first type of normalization is demonstrated by the SAW device array example presented in the previous section. At every point in time, each SAW device output is scaled (divided by) the output of the reference SAW device (coated with bare quartz). When plotted in two-dimensional principal component space (Figure 1.4-5a), the raw data (shift in resonance frequency) exhibit a strong dependence on concentration; in this particular system, concentration data are superfluous to the primary task of differentiating among the 13 analytes. Scaling each individual sensor output with respect to the reference device (Figure 1.4-5b) provides significant resilience to variations in the array response due to concentration changes. By reducing the influence of concentration, data normalization across the sensor array enables the decision-making model to perform with higher accuracy than with raw data. Normalization of each sensor across time can also improve the ability of the decision-making model to discriminate among chemicals while remaining immune to variations in concentration.

Other techniques for scaling individual sensor signals are available, but are highly dependent on the type of sensing technology and purpose of the decision-making model in the system. In all cases, scaling is performed to reduce the superfluous information provided to the decision-making model. In combination with reduction of the array input space, sensor scaling enables a chemical sensing system to focus its computational capability on the part of the overall input space that is directly relevant to the sensing problem at hand rather than on space that is either unoccupied or not useful in multidimensional chemical space.

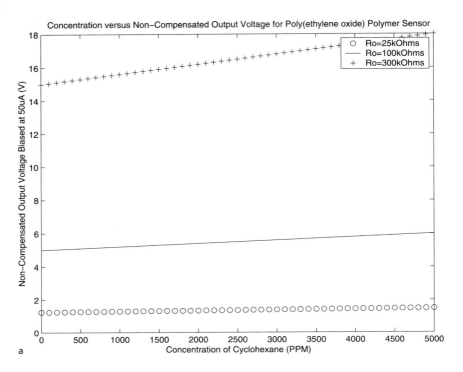

a

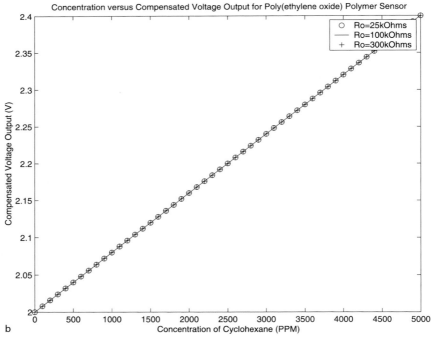

b

1.4.4.4   Feature Extraction

Unlike the reduction of sensor array signals and the scaling of individual sensor signals, feature extraction as a preprocessing technique is very specific to both the type of sensor and the application to which the sensor is applied. Feature extraction requires an understanding of the 'image' provided by an array of sensor data, including both spatial and temporal relationships.

The use of domain-specific knowledge to extract features for transfer to a decision-making model is especially prevalent in image processing. Spatial and temporal relationships are especially intuitive in visual images and for this reason have prompted the expansive and valid use of features of the image rather than individual pixel values to understand, interpret, and process the image for a wide variety of applications. For example, consider the task of differentiating the foreground of an image (the portion of 'interest') and the background of the same image. The detailed content of the foreground in the image may not be relevant or can even be detrimental for high-speed applications such as target tracking. For target tracking, the image can be segmented to determine the edges in the image, smoothed, and otherwise manipulated to reduce the level of detail in the image to a simple binary representation of background and foreground, from which a centroid can be computed and used as input to the control system. In this example, and also many other tasks in image processing, it is important to embed in the feature extraction the ability to detect similar features for the same image, even as background (ambient light) illumination levels change. Image processing provides two valuable lessons for feature extraction:

- Extract features that are suited to the application: background and foreground differentiation are suited to target tracking, whereas edge detection and skeletonization (construction of edges into coherent objects) are more suited to image discrimination.

- Design robustness into the feature extraction that is sufficient to provide resilience to major fluctuations in the sensing environment. In the image processing example, differentiation of foreground and background in a target tracking application must be performed independent of background illumination. It should be irrelevant in determining target location, whether the target is inside or outside, viewed against a forest or blue sky, and so on.

Feature extraction in the preprocessing of chemical sensor signals is an even more complex task than in image processing because intuitive domain-specific

Figure 1.4-6. Preprocessing of chemical sensor information using scaling techniques. The voltage output of the chemiresistor varies (**a**) from 1.25 to 18 V and (**b**) from 2 to 2.4 V before and after baseline compensation, respectively. Reduction of baseline variation results in improved resolution of the system for the same analog-to-digital converter.

knowledge is not as prevalent as it is in the visual field. When chemical sensors are assembled into an array, continuous or semi-continuous relationships among neighboring sensors are rare. The meaning of spatial relationships changes with the composition of the array, often in ways that are not completely understood. In addition, the spatial relationships among neighboring sensors, even when they are understood within their domain, do not vary uniformly across the array. For example, consider an array of metal oxide sensors that are doped with various catalysts and operated at different temperatures. In one dimension of the array, temperature may be varied uniformly across the array (eg, in 10 °C increments). In the other dimension, however, the degree to which a sensor doped with catalyst 1 reacts to analyte 1 may be twice that of that sensor doped with catalyst 2, but 0.75 times that of its other neighboring sensor doped with catalyst 0. To complicate the problem further, the degree of reaction and the change in the reaction of sensors to analytes over both dimensions of the array will, in all likelihood, change with the analyte or combination of analytes present in the sensing environment. Nevertheless, even in the complicated, multidimensional space that chemical sensors provide for interpreting the chemical composition of the sensing environment, features can be extracted following the two basic rules provided by image processing (relevance to the application and robustness to fluctuations in the sensing environment). Sufficient domain-specific knowledge must be incorporated into the extraction of features to validate their use over raw data. Some examples of successful features extracted as preprocessed inputs to decision-making models are discussed and evaluated in the sections that follow.

1.4.4.4.1 Extraction of Electronic Signal from Chemical Sensor

The most basic feature extraction decision in every chemical decision problem focuses on the choice of an electrical parameter to represent the transduction of a chemical signal into electronic form. For chemiresistors, the fundamental electronic output is resistance, which may be extracted as a voltage or current, typically under DC conditions. For a FET-based chemical sensor, work function changes in response to chemical activity cannot be extracted directly but must rely on some measurement of the changing FET threshold voltage. Changing threshold voltage is typically evaluated by maintaining constant current through the FET and measuring resulting voltage or vice versa. Typical measurement techniques involve converting the FET from a three-terminal to a two-terminal device (or four-terminal to three-terminal device) by connecting the drain and gate together to maintain constant current or voltage. Such restriction on the operation of the ChemFET can limit the ability to characterize the electrical performance of the device and should be considered carefully in the measurement circuit design. Acoustic wave devices (surface acoustic wave, bulk acoustic wave, etc.) are similarly complicated in the choice of an electrical parameter for output measurement. The changing velocity of an acoustic wave across these devices can be measured in the most straightforward manner by observing the phase shift in the signal at the input and output terminals of the device. These phase shifts are often so small that alternative techniques such as the measurement of resonant frequency in an oscillator loop must be employed to obtain sufficient sensitivity. Optical sensors typically involve the use of a highly sensitive photodetector to convert optical information (transduced from chemical information) into its final electronic form. Choosing a photodetector involves evaluating another set of output signal extraction issues as well as the matching properties across different detectors used in the same instrument. Electrochemical devices are even more complex because the choice of reference electrode is critical to the stability of the overall system and the magnitude of the voltage or current used to interpret the activity at the working electrode can affect the sensing selectivity of the device.

1.4.4.4.2 Robust Extraction of Electronic Signal from Chemical Sensor

In many cases, the decision as to which electronic signal (resistance, voltage, current, frequency, etc.) is extracted from a chemical signal is based on the data acquisition capability of the resulting measurement instrument. While this form of feature extraction is both the most common and straightforward to impose on chemical sensing systems, the next step in improving such preprocessing is to choose the extracted signal based on its robustness and to match the measurement instrument to the most robust choice to the task at hand. For example, in acoustic wave devices, it may often be more robust to configure the device into an oscillator loop and to evaluate shifts in resonant frequency rather than to evaluate phase shifts in a more straightforward measurement configuration. Although phase shift measurements may be easier in terms of supporting electronics, they do not typically provide comparable detection threshold and sensitivity limits to the resonant frequency measurements. Resonant frequency measurements are also more conducive to modular measurement circuit designs that can be fitted to a variety of SAW device types with minimal additional expense and design effort [38].

1.4.4.4.3 Characteristics of Sensor Transient Response

Such features of a transient event as derivative and length are relevant to the reaction rates between sensor chemistry and sensed analyte; these characteristics can and have been directly applied to chemical discrimination problems including the discrimination of acetone, ethanol, hexane, 2-propanol, methanol, and carbon monoxide [39], acetone, ammonia, denatured alcohol gasoline, motor oil, and xylene [40], and other volatile gases using arrays of tin oxide sensors [41]. In a less straightforward but even more effective manner, temperature pulses have been applied to these same types of sensors to produce concatenated transient responses over temperature that effectively enable metal oxide sensors to discriminate between carbon monoxide and nitric dioxide [42] and among water, ethanol, methanol, formaldehyde, and acetone [43].

1.4.4.4.4 Minimum, Maximum, and Rank Detection

The minimum in certain surface plasmon resonance sensors, after the response curve is scaled to the reference medium (no stimulus presence), is a feature that is directly proportional to the concentration of analyte in the sensing medium (Figure 1.4-7). Details of this portable SPR system can be found in [44, 45]. In most cases, since SPR sensors can be constructed to be highly selective, the location of this minimum in the optical spectrum is the only feature used to calculate analyte concentration. Feature extraction (location of the minimum in the optical spectrum) becomes critical to the decision-making model. Detection of maxima and subsequent ranks have also been demonstrated as useful features in discriminating vapors in metal oxide sensor arrays [46].

1.4.4.4.5 Other Features

A variety of other features have been extracted from arrays of chemical sensors or single chemical sensors in an effort to enhance the concentration detection and analyte discrimination capability of the system. Kato et al. [47] extracted Fourier transform features in the frequency domain from sinusoidally heated metal oxide sensors in order to discriminate ethanol, methanol, diethyl ether, acetone, ethylene, ammonia, isobutane, and benzene at a relatively low concentration level of 100 ppm. Joo et al. [48] selected impedance characteristics from these same types of metal oxide

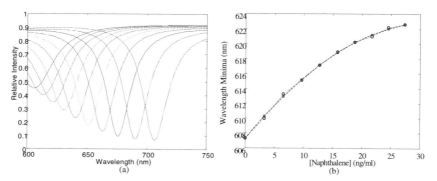

Figure 1.4-7. Preprocessing of chemical sensor information using feature extraction techniques. In a surface plasmon resonance probe, changes in the amount of analyte bound to the surface of the probe result in a change in refractive index of the light reflected along the tip of the probe. This refractive index change is indicated by a shift in the wavelength-dependent light intensity reflected into the measurement detector. After this optical response is scaled by the reference spectrum (produced in the sensing environment with no analytes of interest present), the result is **(a)** a curve whose minimum is directly proportional to the analyte of interest. The extraction of this feature **(b)** (the minimum) provides concentration information directly, eliminating the need for any subsequent signal processing.

sensors by capturing steady-state output voltage and sensor rise and fall time in order to distinguish a wide range of concentrations of C_4H_{10}, CH_4, and carbon monoxide. Other efforts have looked at characteristics of raw data in the full response of sensors including rates of change in polymer sensor resistances [49], time averages and slopes of fiber optic bead sensors [50], and actual reproductions of calibrated sensor odors [51], to name a few.

1.4.5 Opportunities for Future Research

As the demand for more accurate chemical sensing systems increases, the use of array optimization and preprocessing techniques are essential to improving system performance within the constraints of the sensing technology. Efforts to standardize the design of electronic noses and other chemical sensing systems must be taken to the next step where they can be used to design consistently, in modular fashion, chemical sensing systems from the sampling stage to the decision-making phase. Interdisciplinary effort among engineers and chemists is important for bridging the gap between sensing technology and commercially viable sensing systems. As more complex chemical sensing problems demand viable solutions, it is expected that the standardization of systems design will come full circle; knowledge obtained from modular and consistent system design will feed back to the more effective and organized development of sensing technologies for the next generation of chemical sensing systems and products.

1.4.6 Acknowledgments

The authors thank the National Science Foundation for providing partial funding for this research through a pair of CAREER grants (K.S.B., D.M.W.), Jiri Janata at the Georgia Institute of Technology for providing ChemFETs for signal analysis, Richard Cernosek at Sandia National Laboratories for providing the SAW data for analysis; and Nathan Lewis at the California Institute of Technology for providing the composite polymer film data for analysis.

1.4.7 References

[1] Gardner, J.W., Shurmer, H.V., Corcoran, P., *Sens. Actuators B* **4** (1991) 117–121.
[2] Gardner, J.W., Iskandarani, M.Z., Bott, B., *Sens. Actuators B* **9** (1992) 133–142.

[3] Gardner, J. W., *Sens. Actuators B* **27** (1995) 261–266.
[4] Shin, H. W., Lloyd, C., Gardner, J. W., in: *International Conference on Solid-State Sensors and Actuators*, Chicago, IL, 16–19 June, 1997, pp. 935–938.
[5] Shurmer, H. V., Corcoran, P., Gardner, J. W., *Sens. Actuators B* **4** (1991) 29–33.
[6] Craven, M. A., Gardner, J. W., *Trans. Inst. Meas. Control* **20** (1998) 67–73.
[7] Gardner, J. W., Pike, A., de Rooij, N. F., Kouldelka-Hep, M., Clerc, P. A., Hierlemann, A., Göpel, W., *Sens. Actuators B* **26'/27** (1995) 135–139.
[8] Shurmer, H. V., Gardner, J. W., Corcoran, P., *Sens. Actuators* **1** (1990) 256–260.
[9] Gardner, J. W., Shurmer, H. V., Tan, T. T., *Sens. Actuators B* **6** (1992) 71–75.
[10] Sing, S., Hines, E. L., Gardner, J. W., *Sens. Actuators B* **30** (1996) 185– 190.
[11] Harold, V., Gardner, J. W., *Sens. Actuators B* **8** (1992) 1–11.
[12] Llobet, E., Hines, E. L., Gardner, J. W., Franco, S., *Meas. Sci. Technol.* **10** (1999) 538–548.
[13] Gardner, J. W., *Sens. Actuators B* **4** (1991) 109–115.
[14] Gardner, J. W., Pearce, T. C., Friel, S., *Sens. Actuators B* **18/19** (1994) 240–243.
[15] Holmberg, M., Lundstrom, I., Winquist, F., Gardner, J. W., Hines, E. L., Sensors and Actuators, B: Chemical, B27, June 1995, 246–249.
[16] Pike, A., Gardner, J. W., *Sens. Actuators B* **45** (1997) 19–26.
[17] Cavicchi, R. E., Suehle, J. S., Kreider, K. G., Gaitan, M., Chaparala, P., in: *International Conference on Solid-State Sensors and Actuators*, Stockholm, 25–29 June, 1995, pp. 823–826.
[18] Cavicchi, R. E., Semancik, S., Walton, R. M., Panchapakesan, B., DeVoe, D. L., Aquino-Class, M., Allen, J. D., Suehle, J. S., *Proc. SPIE* **3857** (1999) 38–49.
[19] McAvoy, T. J., *Eng. Technol. Sustainable World* **3** (1996) 9–10.
[20] Lewis, N. S., Lonergan, M. C., Severin, E. J., Doleman, B. J., Grubbs, R. H., *Proc. SPIE* **3079** (1997) 660–670.
[21] Lewis, N., in: *IEEE Aerospace Applications Conference,* Snowmass Village, CO, 1–2 February 1997, Vol. 3.
[22] Matzger, A., Vaid, T. P., Lewis, N. S., *Proc. SPIE* **3710** (1999) 315–320.
[23] Ricco, A. J., Crooks, R. M., Osbourn, G. C., *Acc. Chem. Res.* **31** (1998) 289–296.
[24] Cernosek, R. W., Yelton, W. G., Colburn, C. W., Anderson, L. F., Staton, A. W., Osbourn, G. C., Bartholomew, J. W., Martinez, R. F., Ricco, A. J., Crooks, R. M.. *Proc. SPIE* **3857** (1999) 146–157.
[25] Osbourn, G. C., Martinez, R. F., *Pattern Recognit.* **28** (1995) 1793–1806.
[26] Wong, C. C., Adkins, D. R., Frye-Mason, G. C., Hudson, M. L., Kottenstette, R., Matzke, C. M., Shadid, J. N., Salinger, A. G., *Proc. SPIE* **3877** (1999) 120–129.
[27] Baca, A. G., Heller, E. J., Hietala, V. M., Casalnuovo, S. A., Frye-Mason, G. C., Klem, J. F., Drummon, T. J., *IEEE J. Solid State Circuits* **34** (1999) 1254–1258.
[28] Thomas, G. A., Frye-Mason, G. C., Bailey, C., Warren, M. E., Fruetel, J. A., Wally, K., Wu, J., Kottenstette, R. J., Heller, E. J., *Proc. SPIE* **3713** (1999) 66–76.
[29] Wilson, D. M., *Electron. Lett.* **32** (1996) 991–992.
[30] Wilson, D. M., Roppel, T., in: *SPIE Proceedings Photonics East*, Boston, MA, 20–22 September, **3856** (1999) 171–180.
[31] Lewis, N. S., Lonergan, M. C., Severin, E. J., Doleman, B. J., Grubbs, R. H., *Proc. SPIE* **3079** (1997) 660–670.
[32] Gardner, J. W., Bartlett, P. N., *Sens. Actuators B* **33** (1996) 60–67.
[33] Neibling, G., Muller, R., *Sens. Actuators B* **25** (1995) 3, 781–784.
[34] Chaurdy, A. N., Hawkins, T. M., Travers, P. J., *Sens. Actuators B* **69** (2000) 236–242.
[35] Gray, P. R., Meyer, R. G., *Analysis and Design of Analog Integrated Circuits*; New York: Wiley, 1977, pp. 635–702.

[36] Kish, L. B., Vajtai, R., Granqvist, C. G., *Sens. Actuators B* **71** (2000) 55–59.
[37] Kennedy, R. L., Lee, Y., Van Roy, B., Reed, C. D., Lippmann, R. P., *Solving Data Mining Problems Through Pattern Recognition*; Englewood Cliffs, NJ: Prentice Hall, 1995–97, Ch. 9.
[38] Schmitt, R. F., Allen, J. W., Wright, R., *Sens. Actuators B* **76** (2001) 80–85.
[39] Wilson, D. M., DeWeerth, S. P., *Sens. Actuators B*, **28** (1995) 123–128.
[40] Roppel, T., Dunman, K., Padgett, M., Wilson, D., Lindbald, T., in: *Proceedings of the IECON*, 1997, Vol. 1, pp. 218–221.
[41] Hiranaka, Y., Abe, T., in: *1991 International Conference on Solid-State Sensors and Actuators*; San Francisco, 24–28 June 1991, pp. 157–160.
[42] Jaegle, M., Woellenstein, J., Meisinger, T., Boettner, H., Mueller, G., Becker, T., Bosch v. Braunmuehl, C., *Sens. Actuators B*, **57** (1999) 130–134.
[43] Cavichhi, R. E., Suehle, J. S., Kreider, K. G., Gaitan, M., Chaparala, P., *IEEE Electron. Device Lett.* **16** (1995) 286–288.
[44] Obando, L. A., Booksh, K. S., *Anal. Chem.* **71** (1999) 5116–5122.
[45] Boysworth, M. K., Obando, L. A., Booksh, K. S., *Proc. SPIE* **3854** (1999) 308–316.
[46] Wilson, D. M., Dunman, K., Roppel, T., Kalim, R., *Sens. Actuators B* **62** (2000) 199–210.
[47] Kato, K., Kato, Y., Takamatsu, K., Udaka, T., Nakahara, T., Matsuura, Y., Yoshikawa, K., *Sens. Actuators B* **71** (2000) 192–196.
[48] Joo, B.-S., Choi, N.-J., Lee, Y.-S., Lim, J.-W., Kang, B.-H., Lee, D.-D., *Sens. Actuators B* **77** (2001) 209–214.
[49] Magan, N., Pavlou, A., Chrysanthakis, I., *Sens. Actuators B* **72** (2001) 28–34.
[50] Bakken, G. A., Kauffman, G. W., Jurs, P. C., Albert, K. J., Stitzel, S. S., *Sens. Actuators B* **79** (2001) 1–10.
[51] Nakamoto, T., Nakahira, Y., Hiramatsu, H., Moriizumi, T., *Sens. Actuators B* **76** (2001) 465–469.

List of Symbols and Abbreviations

Symbol	Designation
f	frequency
i	sensor
i	first cluster
j	second cluster
j	sensor
k	experiment
M	number of heterogeneous sensors
N	number of experiments to optimize an array
N	number of sensors in an array
P	partial pressure
P_{sat}	saturated partial pressure
R_i	signal

Symbol	Designation
R_{jk}	response of sensor j during experiment
RP	resolving power
μ_{ik}	mean of sensor k response for the first cluster
$\mu_{d(i,j)}$	average distance from first cluster mean i to second cluster members j
$\sigma_{d(i,j)}$	variance in distances between first and second clusters
σ^2	variance

Abbreviation	Explanation
ADC	analog-to-digital converter
ANN	artificial neural network
ChemFET	chemical field effect transistor
DIMP	diisopropyl methylphosphonate
DMMP	dimethyl methylphosphonate
FET	field effect transistor
NIST	National Institute of Standards and Technology
SAW	surface acoustic wave
TCE	trichloroethylene
THF	tetrahydrofuran

1.5 Glucose Biosensors: 40 Years of Advances and Challenges

J. WANG, Department of Chemistry and Biochemistry,
New Mexico State University, Las Cruces, NM 88003, USA

Abstract

Forty years have passed since Clark and Lyons proposed the concept of glucose enzyme electrodes. Excellent economic prospects and fascinating potential for basic research have led to many sensor designs and detection principles for the biosensing of glucose. Indeed, the entire field of biosensors can trace its origin to this glucose enzyme electrode. This review examines the history of electrochemical glucose biosensors, discusses their current status and assesses future prospects in connection primarily to the control and management of diabetes.

Keywords: Glucose; biosensor; diabetes; enzyme electrodes

Contents

1.5.1   Introduction

Diabetes is a world-wide public health problem. It is one of the leading causes of death and disability in the world. The diagnosis and management of diabetes mellitus requires a tight monitoring of blood glucose levels. The challenge of providing such tight and reliable glycemic control remains the subject of enormous amount of research [1, 2]. Electrochemical biosensors for glucose play a leading role in this direction. Amperometric enzyme electrodes, based on glucose oxidase (GOx) bound to electrode transducers, have thus been the target of substantial research [1, 2].

Since Clark and Lyons first proposed the initial concept of glucose enzyme electrodes in 1962 [3] we have witnessed tremendous activity towards the development of reliable devices for diabetes control. A variety of approaches have been explored in the operation of glucose enzyme electrodes. In addition to diabetes control, such devices offer great promise for other important applications, ranging from food analysis to bioprocess monitoring. The great importance of glucose has generated an enormous number of publications, the flow of which shows no sign of diminishing. Yet, despite of impressive advances in glucose biosensors, there are still many challenges related to the achievement of clinically accurate tight glycemic monitoring.

The goal of this review article is to examine the history of electrochemical glucose biosensors, assess their current status, and discuss future challenges.

1.5.2   Forty Years of Progress

The idea of a glucose enzyme electrode was proposed in 1962 by Clark and Lyons from the Children Hospital in Cincinnati [3]. Their first device relied on a thin layer of GOx entrapped over an oxygen electrode (via a semipermeable dialysis membrane), and monitoring the oxygen consumed by the enzyme-catalyzed reaction:

$$\text{glucose} + \text{oxygen} \xrightarrow{\text{GOx}} \text{gluconic acid} + \text{hydrogen peroxide} \qquad (1)$$

Clark's original patent [4] covers the use of one or more enzymes for converting electroinactive substrates to electroactive products. The effect of interferences was corrected by using two electrodes (one covered with GOx) and measuring the differential current. Clark's technology was subsequently transferred to Yellow Spring Instrument Company that launched in 1975 the first dedicated glucose analyzer (the Model 23 YSI analyzer) for the direct measurement of glucose in 25 µL samples of whole blood. Updike and Hicks [5] further developed this principle by using two oxygen working electrodes (one covered with the enzyme) and measuring the differential current for correcting for the oxygen background variation in samples. Guilbault and Lubrano [6] described in 1973 an enzyme electrode for the determination of blood glucose based on amperometric (anodic) monitoring of the liberated hydrogen peroxide:

$$H_2O_2 \rightarrow O_2 + 2H^+ + 2e^- \tag{2}$$

Good precision and accuracy were obtained in connection to 100 µL blood samples. A wide range of amperometric enzyme electrodes, differing in the electrode design or material, membrane composition, or immobilization approach have since been described.

During the 1980s biosensors became a 'hot' topic, reflecting the growing emphasis on biotechnology. Intense efforts during this decade focused on the development of mediator-based 'second-generation' glucose biosensors [7, 8], the introduction of commercial strips for self-monitoring of blood glucose [9, 10], and the use of modified electrodes for enhancing the sensor performance [11]. In the 1990s we witnessed intense activity towards the establishment of electrical communication between the GOx redox center and the electrode surface [12, 13], and the development of minimally-invasive subcutaneously implantable devices [14–16]. Table 1.5-1 summarizes major historical landmarks in the development of electrochemical glucose biosensors.

Table 1.5-1. Historical landmarks in the development of electrochemical glucose biosensors.

Date	Event	References
1962	First glucose enzyme electrode	[3]
1973	Glucose enzyme electrode based on peroxide detection	[6]
1975	Launch of the first commercial glucose sensor system	YSI Inc.
1982	Demonstration of in vivo glucose monitoring	[31]
1984	Development of ferrocene mediators	[7]
1987	Launch of the first personal glucose meter	Medisense Inc.
1987	Electrical wiring of enzymes	[12a]
1999	Launch of a commercial in vivo glucose sensor	Minimed Inc.
2000	Introduction of a wearable noninvasive glucose monitor	Cygnus Inc.

1.5.3   First-Generation Glucose Biosensors

First-generation devices have relied on the use of the natural oxygen cosubstrate, and the production and detection of hydrogen peroxide (Equations 1 and 2). Such measurements of peroxide formation has the advantage of being simpler, especially when miniaturized sensors are concerned. A very common configuration is the YSI probe, involving the entrapment of GOx between an outer diffusion-limiting/biocompatible polycarbonate membrane and an inner anti-interference cellulose acetate one (Figure 1.5-1).

1.5.3.1   Redox Interferences

The amperometric measurement of hydrogen peroxide requires application of a potential at which coexisting species, such as ascorbic and uric acids or acetaminophen, are also electroactive. The anodic contributions of these and other oxidizable constituents of biological fluids can compromise the selectivity and hence the overall accuracy. Extensive efforts during the 1980s were devoted for minimizing the error of electroactive interferences in glucose electrodes. One useful strategy is to employ a permselective coating that minimizes access of such constituents to the transducer surface. Different polymers, multilayers and mixed layers, with transport properties based on size, charge, or polarity, have thus been used for discriminating against coexisting electroactive compounds [17–19]. Such films also exclude surface-active macromolecules, hence imparting higher stability. Electropolymerized films, particularly poly(phenylenediamine) and overoxidized polypyrrole, have been shown particularly useful for imparting high selectivity (based on size exclusion) while confining the GOx onto the surface [17, 18]. Other widely used coatings include the negatively charged (sulfonated) Nafion or Kodak AQ ionomers, size-exclusion cellulose acetate films, and hydrophobic alkanethiol or lipid layers. The use of multi-(overlaid) layers, com-

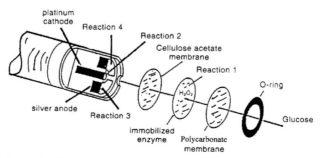

Figure 1.5-1. Schematic of a "first-generation" glucose biosensor (based on a probe manufactured by YSI Inc.).

bining the properties of different films, offers additional advantages. For example, alternate deposition of cellulose acetate and Nafion was used for eliminating the interference of the neutral acetaminophen and negatively charged ascorbic and uric acids, respectively [19].

Efforts during the 1990s focused on the preferential electrocatalytic detection of the liberated hydrogen peroxide [20–23]. This has allowed tuning of the detection potential to the optimal region (ca. 0.0 to –0.20 V vs. Ag/AgCl) where most unwanted background reactions are negligible. The remarkably high selectivity thus obtained was coupled to a fast and sensitive response. Metalized (Rh,Ru)-carbon [20, 21] and metalhexacyanoferrate [22, 23] based transducers have been particularly useful for enhancing the selectivity towards the target glucose substrate. Additional improvements can be achieved by combining this preferential catalytic activity with a discriminative layer, e.g., by dispersing rhodium particles within a Nafion film [24].

1.5.3.2 Oxygen Dependence

Since oxidase-based devices rely on the use of oxygen as the physiological electron acceptor, they are subject to errors accrued from fluctuations in the oxygen tension and the stoichiometric limitation of oxygen. Such limitation (known as the "oxygen deficit") reflects the fact that normal oxygen concentrations are about an order of magnitude lower than the physiological level of glucose.

Several routes have been proposed for addressing this oxygen limitation. One strategy relies on the use of mass-transport limiting films (such as polyurethane or polycarbonate) for tailoring the flux of glucose and oxygen, i.e., increasing the oxygen/glucose permeability ratio [1, 25]. A two-dimensional electrode, designed by Gough's group [25], has been particularly attractive for addressing the oxygen deficit by allowing oxygen to diffuse into the enzyme region of the sensor from both directions and glucose diffusion only from one direction. We have recently addressed the oxygen limitation of glucose biosensors by designing an oxygen-rich carbon paste enzyme electrode [26]. The new biosensor is based on a fluorocarbon (Kel-F oil) pasting liquid, which has very high oxygen solubility, allowing it to act as an internal source of oxygen. The internal flux of oxygen can thus support the enzymatic reaction even in oxygen-free glucose solutions. It is possible also to circumvent the oxygen demand issue by replacing the GOx with glucose dehydrogenase (GDH) that does not require an oxygen cofactor [27].

1.5.4   Second-Generation Glucose Biosensors

1.5.4.1   Electron Transfer between GOx and Electrode Surfaces

Further improvements (and attention to the above errors) can be achieved by replacing the oxygen with a nonphysiological (synthetic) electron acceptor, which
is able to shuttle electrons from the redox center of the enzyme to the surface of
the electrode. Glucose oxidase does not directly transfer electrons to conventional electrodes because a thick protein layer surrounds its flavin redox center.
Such thick protein shell introduces a spatial separation of the electron donor-ac-
ceptor pair, and hence an intrinsic barrier to direct electron transfer, in accor-
dance to the distance dependence of the electron transfer rate [28]:

$$K_{et} = 10^{13} e^{-0.91(d-3)} e^{[-(\Delta G + \lambda)/4RT\lambda]} \tag{3}$$

where ΔG and λ correspond to the free and reorganization energies accompany-
ing the electron transfer, respectively, and d the actual electron transfer distance.
The minimization of the electron-transfer distance (between the immobilized
GOx and the electrode surface) is thus crucial for ensuring optimal performance.
Accordingly, various innovative strategies have been suggested for establishing
and tailoring the electrical contact between the redox center of GOx and electrode surfaces.

1.5.4.2   Use of Artificial Mediators

Particularly useful has been the use of artificial mediators that shuttle electrons
between the FAD center and the surface by the following scheme:

$$glucose + GOx_{(ox)} \; \rightarrow \; gluconic \; acid + GOx_{(red)} \tag{4}$$

$$GOx_{(red)} + 2M_{(ox)} \; \rightarrow \; GOx_{(ox)} + 2M_{(red)} + 2H^+ \tag{5}$$

$$2M_{(red)} \; \rightarrow \; 2M_{(ox)} + 2e^- \tag{6}$$

where $M_{(ox)}$ and $M_{(red)}$ are the oxidized and reduced forms of the mediator. Such
mediation cycle produces a current dependent on the glucose concentration.
Diffusional electron mediators, such as ferrocene derivatives, ferricyanide, conducting organic salts (particularly tetrathiofulvalene-tetracyanoquinodimethane,
TTF-TCNQ), phenothiazine and phenoxazine compounds, or quinone compounds
have thus been widely used to electrically contact GOx [7, 8] (Figure 1.5-2). As
a result of using these electron-carrying mediators, measurements become largely
independent of oxygen partial pressure and can be carried out at lower potentials

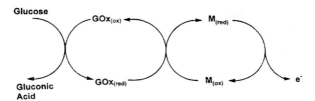

Figure 1.5-2. Sequence of events that occur in 'second-generation' (mediator-based) glucose biosensors-mediated system.

that do not provoke interfering reactions from coexisting electroactive species (Equation 6). In order to function effectively, the mediator should react rapidly with the reduced enzyme (to minimize competition with oxygen), possess good electrochemical properties (such as a low redox potential), have low solubility in aqueous medium, and must be nontoxic and chemically stable (in both reduced and oxidized forms). Commercial blood glucose self-testing meters, described in the following section, commonly rely on the use of ferrocene or ferricyanide mediators. Most in vivo devices, however, are mediatorless due to potential leaching and toxicity of the mediator.

1.5.4.3   Attachment of Electron-Transfer Relays

Heller's group [12] developed an elegant nondiffusional route for establishing a communication link between GOx and electrodes based on 'wiring' the enzyme to the surface with a long flexible poly-pyridine polymer having a dense array of osmium complex electron relays. The resulting three-dimensional redox polymer/ enzyme networks offer high current outputs and stabilize the mediator to electrode surfaces.

Chemical modification of GOx with electron-relay groups represents another novel avenue for facilitating the electron transfer between its redox center and the electrode surface. Willner and co-workers [13] reported on an elegant approach for modifying GOx with electron relays (Figure 1.5-3). For this purpose, the FAD active center of the enzyme was removed to allow positioning of an electron-mediating ferrocene unit prior to the reconstitution of the enzyme. The attachment of electron-transfer relays at the enzyme periphery has also been considered for yielding short electron-transfer distances [29]. More sophisticated bioelectronic systems for enhancing the electrical response, based on patterned monolayer or multilayer assemblies and organized enzyme networks on solid electrodes, have been developed for contacting GOx with the electrode support [29]. Functionalized alkanethiol-modified gold surfaces have been   particularly attractive for such layer-by-layer creation of GOx-mediator networks.

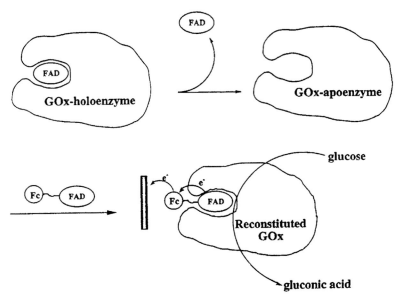

Figure 1.5-3. Electrical contacting of a flavoenzyme by its reconstitution with a relay-FAD semisynthetic cofactor (reproduced with permission [29b]).

1.5.5   In Vitro Glucose Testing

Electrochemical biosensors are well suited for satisfying the needs of home (personal) glucose testing. The majority of personal blood glucose meters are based on disposable (screen-printed) enzyme electrode test strips. Such single-use disposable electrode strips are mass produced by the thick-film (screen-printing) microfabrication technology. Each strip contains the printed working and reference electrodes, with the working one coated with the necessary reagents (i.e., enzyme, mediator, stabilizer, linking agent). Such reagents are commonly being dispensed by an ink-jet printing technology. A counter and an additional ('baseline') working electrode may also be included. Such single-use devices obviate problems of carry over, contamination, or drift.

The control meter is typically light and small (pocket-size), battery operated, and relies on a potential-step (chronoamperometric) operation. Such devices offer considerable promise for obtaining the desired clinical information in a faster, simpler ("user-friendly"), and cheaper manner compared to traditional assays. The first product was a pen-style device (the Exactech), launched by Medisense Inc. in 1987, that relied on the use of a ferrocene-derivative mediator. Various commercial strips and pocket-sized test meters, for self-monitoring of blood glucose based on the use of ferricyanide or ferrocene mediators – have since been introduced (Table 1.5-2) [30]. In most cases, the diabetic patient pricks the fin-

Table 1.5-2. Commercial electrochemical systems for self-monitoring of blood glucose.

Source	Trade name	Enzyme	Mediator
Abbot/Medisense	Precision	GOx	Ferrocene
Bayer	Elite	GOx	Ferricyanide
LifeScan	SureStep	GOx	Ferricyanide
Roche-Diagnostics	Accu-Check	GOx	Ferricyanide
Therasense	FreeStyle	GOx	Osmium 'wire'

ger, places the small blood droplet on the sensor strip, and obtains the blood glucose concentration (on a LC display) within 15–30 s. Recent efforts have led to new strips, requiring sub-micrometer blood volumes and enabling "less painful" sampling from the arm. In addition to small size, fast response, and minimal sample requirements, such modern personal glucose meters have features such as extended memory capacity and computer downloading capabilities.

1.5.6   Continuous In Vivo Monitoring

Although self testing is considered a major advance in glucose monitoring it is limited by the number of tests per a 24 h period. Such testing neglect nighttime variations and may result in poor approximation of blood glucose variations. Tighter glycemic control, through more frequent measurements or continuous monitoring, is desired for triggering proper alarm in cases of hypo- and hyperglycemia, and for making valid therapeutic decisions [15]. A wide range of possible in vivo glucose biosensors has thus been studied for maintaining glucose levels close to normal. The first application of such devices for in vivo glucose monitoring was demonstrated first by Shichiri et al. in 1982 [31]. Continuous ex vivo monitoring of blood glucose was proposed already in 1974 [32].

1.5.6.1   Requirements

The major requirements of clinically accurate in vivo glucose sensors have been discussed in various review articles [1, 15]. These include proper attention to the issues of biocompatibility/biofouling, miniaturization, long-term stability of the enzyme and transducer, oxygen deficit, baseline drift, short stabilization times, in vivo calibration, safety, and convenience. The sensor must be of a size and shape that can be easily implanted and cause minimal discomfort. Under biocompatibility one must consider the effect of the sensor upon the in vivo environment as well as the environment effect upon the sensor performance. Problems with biocompatibility have proved to be the major barriers to the development of reliable implantable devices. Most glucose biosensors lack the biocompatibility necessary for a prolonged and reliable operation in whole blood. Alternative sensing sites, particularly the subcutaneous tissue, have thus received growing attention. While the above issues represent a major challenge, significant progress has been made towards the continuous monitoring of glucose.

1.5.6.2   Subcutaneous Monitoring

Most of the recent attention has been given to the development of subcutaneously implantable needle-type electrodes (Figure 1.5-4) [14–16]. Such devices are designed to operate for a few days and be replaced by the patient. Success in this direction has reached the level of short-term human implantation; continuously functioning devices, possessing adequate (>1 week) stability, are expected in the near future. Such devices would enable a swift and appropriate corrective action (through a closed-loop insulin delivery system, i.e., an artificial pancreas). Algorithms correcting for the transient difference (time lag) between blood and tissue glucose concentrations have been developed [16]. The recently introduced CGMS unit of Minimed Inc. (Sylmar, CA) offers a 72 h of such subcutaneous monitoring, with measurement of tissue glucose every 5 min and data storage in the monitor's memory. After 72 h, the sensor is removed, and the information is transferred to a computer for identifying patterns of glucose variations. In addition to easily removable short-term implants, efforts are continuing towards chronically implanted devices (aimed at functioning reliably 6–12 months).

1.5.6.3   Towards Noninvasive Glucose Monitoring

Noninvasive approaches for continuous glucose monitoring represent a promising route for obviating the challenges of implantable devices. In particular, Cygnus Inc. has developed an attractive wearable glucose monitor, based on the coupling

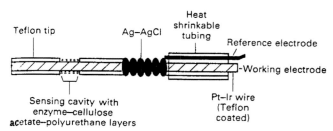

Figure 1.5-4. A needle-type glucose biosensor for subcutaneous monitoring (reproduced with permission [14]).

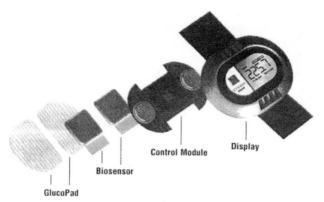

Control Module   Display

Biosensor

GlucoPad

Figure 1.5-5. Various components of the watch-like glucose monitor being developed by Cygnus Inc.

of reverse iontophoretic collection of glucose and biosensor functions [33]. The new GlucoWatch biographer (shown in Figure 1.5-5) provides up to three glucose readings per hour for up to 12 h (i.e., 36 readings within a 12 h period). The system has been shown to be capable of measuring the electroosmotically extracted glucose with a clinically acceptable level of accuracy. An alarm capability is included to alert the individual of very low or high glucose levels. Other routes for "collecting" the glucose through the skin and for noninvasive glucose testing are currently being examined by various groups and companies.

1.5.7   Conclusions and Prospects

Over the past forty years we have witnessed an intense activity and tremendous progress towards the development of electrochemical glucose biosensors. Major advances have been made for enhancing the capabilities and improving the reliability of glucose measuring devices. Such intensive activity has been attributed to tremendous economic prospects and fascinating research opportunities. The success of glucose blood monitors has stimulated considerable interest in in vitro and in vivo devices for monitoring other physiologically important compounds. Despite the impressive advances in glucose biosensors, there are still many challenges related to the achievement of tight, stable and reliable glycemic monitoring. The development of new and improved glucose biosensors thus remains the prime focus of many researchers.

As this field enters the fifth decade of intense research we expect significant efforts coupling fundamental sciences with technological advances. Such stretching of the ingenuity of researchers will result in greatly improved electrical contact between the redox center of GOx and electrode surfaces, enhanced "genetically engineered" GOx, new "painless" in vitro testing, advanced biocompatible membrane materials, the coupling of minimally invasive monitoring with compact insulin delivery system, new innovative approaches for noninvasive monitoring, and miniaturized long-term implants. These, and similar developments, will greatly improve the control and management of diabetes.

1.5.8   Acknowledgement

Financial support from the National Institute of Health (NIH grant RO1 RR 14173-03) is gratefully acknowledged.

1.5.9   References

[1]   Reach, G., Wilson, G.S., *Anal. Chem.* **64** (1992) 381A.
[2]   Turner, A.P., Chen, B., Piletsky, S., *Clin. Chem.* **45** (1999) 1596.
[3]   Clark, L. Jr., Lyons, C., *Ann. NY Acad. Sci.* **102** (1962) 29.
[4]   Clark, L. Jr., *US Patent 33 539 455*, 1970.
[5]   Updike, S., Hicks, G., *Nature* **214** (1967) 986.
[6]   Guilbault, G., Lubrano, G., *Anal. Chim. Acta* **64** (1973) 439.
[7]   Cass, A., Davis, G., Francis, G., Hill, H.A., Aston, W., Higgins, I.J., Plotkin, E., Scott, L., Turner, A.P., *Anal. Chem.* **56** (1984) 667.
[8]   Frew, J., Hill, H.A., *Anal. Chem.* **59** (1987) 933A.
[9]   Hilditch, P., Green, M., *Analyst* **116** (1991) 1217.
[10]   Matthews, D., Holman, R., Brown, E., Streemson, J., Watson, A., Hughes, S., *Lancet* **2** (1987) 778.
[11]   Murray, R.W., Ewing, A., Durst, R., *Anal. Chem.* **59** (1987) 379A.
[12]   a) Degani, Y., Heller, A., *J. Phys. Chem.* **91** (1987) 1285; b) Ohara, T., Rajogopalan, R., Heller, A., *Anal. Chem.* **66** (1994) 2451.
[13]   Willner, I., Heleg-Shabtai, V., Blonder, R., Katz, E., Tao, G., *J. Am. Chem. Soc.* **118** (1996) 10321.
[14]   Bindra, D., Zhang, Y., Wilson, G., Sternberg, R., Trevenot, D., Reach, G., Moatti, D., *Anal. Chem.* **63** (1991) 1692.
[15]   Henry, C., *Anal. Chem.* **70** (1998) 594A.
[16]   Schmidtke, D., Freeland, A., Heller, A., Bonnecaze, R., *Proc. Natl. Acad. Sci. USA* **95** (1998) 294.
[17]   a) Sasso, S., Pierce, R., Walla, R., Yacynych, A., *Anal. Chem.* **62** (1990) 1111; b) Emr, S., Yacynych, A., *Electroanalysis* **7** (1995) 913.
[18]   Malitesta, C., Palmisano, F., Torsi, L., Zambonin, P., *Anal. Chem.* **62** (1990) 2735.
[19]   Zhang, Y., Hu, Y., Wilson, G.S., Moatti-Sirat, D., Poitout, V., Reach, G., *Anal. Chem.* **66** (1994) 1183.
[20]   Wang, J., Liu, J., Chen, L., Lu, F., *Anal. Chem.* **66** (1994) 3600.
[21]   Newman, J., White, S., Tothill, I., Turner, A.P., *Anal. Chem.* **67** (1995) 4594.

[22] Karaykin, A., Gitelmacher, O., Karaykina, E., *Anal. Chem.* **67** (1995) 2419.

[23] Chi, Q., Dong, S., *Anal. Chim. Acta* **310** (1995) 429.

[24] Wang, J., Wu, H., *J. Electroanal. Chem.* **395** (1995) 287.

[25] a) Gough, D., Lucisano, J., Tse, P., *Anal. Chem.* **57** (1985) 2351; b) Armour, J., Lucisano, J., Gough, D., *Diabetes* **39** (1990) 1519.

[26] Wang, J., Lu, F., *J. Am. Chem. Soc.* **120** (1998) 1048.

[27] a) D'Costa, E., Higgins, I., Turner, A.P., *Biosensors* **2** (1986) 71; b) Laurinavicius, V., Kurtinaitiene, B., Liauksminas, V., Ramanavicius, A., Meskys, R., Rudomanskis, R., Skotheim, T., Boguslasky, L., *Anal. Lett.* **32** (1999) 299.

[28] Marcus, R.A., Sutin, N., *Biochim. Biophys. Acta* **811** (1985) 265.

[29] a) Willner, I., Katz, E., *Angew. Chem. Int. Ed.* **39** (2000) 1180; b) Willner, I., Katz, E., Willner, B., *Electroanalysis* **7** (1995) 23; c) Hodak, J., Etchenique, R., Calvo, E., Bartlett, P.N., *Langmuir* **13** (1997) 2708.

[30] Kirk, J., Rheney, C., *J. Am. Pharm. Assoc.* **38** (1998) 210.

[31] Shichiri, M., Yamasaki, Y., Hakui, N., Abe, H., *Lancet* **2** (1982) 1129.

[32] Albisser, A., Lebel, B., Ewart, G., Davidovac, Z., Botz, C., Zingg, W., *Diabetes* **23** (1974) 397.

[33] Tierney, M., Kim, H., Tamada, J., Potts, R., *Electroanalysis* **12** (2000) 666.

PART 2

Sensor Applications

2.1 Enantioselective Electrochemical Sensors

R.-I. STEFAN, University of Pretoria, Pretoria, South Africa
H.Y. ABOUL-ENEIN, King Faisal Specialist Hospital
and Research Centre, Riyadh, Saudi Arabia
J. F. VAN STADEN, University of Pretoria, Pretoria, South Africa

Abstract

The discrimination between enantiomers is becoming one of the most important fields of modern analytical chemistry, especially for pharmaceutical products, clinical analysis, and food analysis. The introduction of enantioselective electrochemical sensors in enantioanalysis increases the reliability of the measurements. Three types of enantioselective electrochemical sensors are proposed for enantioselective analysis: potentiometric, enantioselective electrodes, amperometric immunosensors, and amperometric biosensors. The principle of enantiomer recognition is the same for all types of proposed electrochemical sensors: to find a lock for the key, where the key is the enantiomer, and the lock is a molecule with a special architecture, an antibody, or an enzyme.

Keywords: Enantioselective sensors; potentiometry; amperometry; biosensor; immunosensor

Contents

2.1.1   Introduction

Nowadays, enantioselective analysis is becoming increasingly important especially for the pharmaceutical industry and as part of clinical analysis and food
analysis. As most body substances are built with amino acids chains, for pharmaceutical compounds having a chiral center only one of the enantiomers will be
biologically active. The other enantiomer can be inactive, toxic, or with a different biological activity causing unexpected secondary effects [1].

The introduction of enantioselective electrochemical sensors increased the reliability of the assay of the enantiomers because the enantiomer can be determined
without prior separation, directly from the matrix, only dissolution and dilution
steps being involved [2–7]. The time of analysis is short. The precision of the assay of the enantiomer is high.

Furthermore, most enantioselective sensors can be used as detectors in flow systems. On-line enantioanalysis with electrochemical sensors was designed for the assay of only one of the enantiomers [8–10], or for the simultaneous assay of both enantiomers [11, 12].

The selection of the type of enantioselective sensor for on-line enantioanalysis must take into account the concentration of the enantiomer in the solution, eg, for pharmaceutical compounds potentiometric, the enantioselective membrane electrodes is preferred for the determination of the main enantiomer [11] whereas the amperometric biosensor is preferred for the assay of the other enantiomer which is present in low concentration [11, 12].

Flow injection analysis (FIA) [8, 9] and sequential injection analysis (SIA) [10–12] have both been applied for enantioselective analysis with enantioselective sensors as detectors. As SIA involves smaller quantities of sample and buffer and a cheap carrier (NaCl solution, 0.1 or 1.0 mol/L), its application for enantioselective analysis is preferred to FIA.

2.1.2 Classification of Enantioselective Electrochemical Sensors

There are three main classes of electrochemical sensors that are used in enantioselective analysis: potentiometric, enantioselective membrane electrodes, amperometric immunosensors, and amperometric biosensors. The selection of the class of sensor to be used for enantioselective analysis must take into account the complexity of the chiral molecule and also the quantity of the enantiomer to be determined [4]. Up to now, amperometric immunosensors and amperometric biosensors provided very good accuracy and precision for enantiomers of complex molecules, and potentiometric, enantioselective membrane electrodes for enantiomers of simple molecules.

Another classification of enantioselective electrochemical sensors can be made according to their design. There are two main classes of enantioselective electrochemical sensors: carbon paste electrodes and plastic membrane-based sensors. The main advantage of the utilization of carbon paste electrodes is the high reliability of membrane design, which will give highly reliable response characteristics to the electrode.

2.1.3 Principles

There is only one main principle for enantioselective analysis using electrochemical sensors: *to find the lock for the key*. The key represents the enantiomer that must be analyzed. For the lock, there are two possibilities:

(i) a substance with a special architecture that can bind enantioselectively with the enantiomer;

(ii) an enzyme that can only catalyze the reaction of the main enantiomer.

Accordingly, these principles can be defined as:

(i) enantioselective binding;

(ii) enantioselective catalysis.

2.1.3.1   Enantioselective Binding

This principle is applied for potential development in potentiometric, enantioselective membrane electrodes, and for obtaining the intensity of the current in amperometric immunosensors. For potentiometric, enantioselective electrodes it is necessary to find a molecule with a special architecture that can accommodate the enantiomer. In this regard, cyclodextrin derivatives, maltodextrins, and quinine and quinidine derivatives have been proposed [2–7].

The advantage of using cyclodextrins and maltodextrins as chiral selectors is that the enantioselectivity of the molecular interaction takes place in two places: inside the cavity (internal enantioselectivity) and outside the cavity – due to the arrangement, size, and type of the radicals, atoms, or ions bound on the external chain of the chiral selector (external enantioselectivity). The thermodynamics of the reaction between the enantiomers and chiral selectors plays the main role in the enantioselectivity of molecular interaction.

Immunosensors are based on the reaction between antigen (Ag) and antibody (Ab). Each enantiomer (as Ag) will have its pair Ab. That is why the enantioselectivity of the reaction between Ag and Ab is the highest between the enantioselective sensors. If this is also coupled with an amperometric transducer, the immunosensor will also be characterized by high sensitivity. The decrease in sensitivity can be recorded when the complex formed between Ag and Ab is not electrochemically active, and the Ab must be coupled with an enzyme (enzyme-linked immunosorbent assay (ELISA) technique). Through the coupling reaction the reactivity of each compound is modified, and therefore it can also be decreased for the given Ab, and, as a result, the sensitivity of the amperometric immunosensor will also decrease, and also the accuracy of the measurements, as the determination of the enantiomer will not be a direct one.

2.1.3.2   Enantioselective Catalysis

The biological medium is a very valuable source of enzymes that can successfully be used for enantioselective catalysis. If one is looking at the processes that take place in the body or in plants, it is easy to see that most of them need a catalyst that is a specific enzyme. The enzymes can be isolated and purified and characterized through their activity before and after using in the design of the biosensor.

The source of the enzyme is very important. The same type of enzyme isolated from different media can assure different response characteristics in terms of sensitivity and limits of detection.

By coupling the high enantioselectivity of the enzyme with the high sensitivity of the amperometric transducer, the enantioselective amperometric biosensor can be used successfully for the assay of the most complex chiral substances.

Owing to the high importance of the assay of the enantiomers of amino acids and amino acids derivatives, the assay of the enantiomers of these substances can be done by using enantioselective amperometric biosensors based on L-amino acid oxidase (L-AAOD) and D-amino acid oxidase (D-AAOD) for the assay of *S*- and *R*-enantiomers, respectively.

2.1.4 Mechanism of Potential Development and Enantioselectivity for Potentiometric, Enantioselective Membrane Electrodes

The main processes that are taking place at the membrane-solution interface are as follows [13]:

1. extraction of the enantiomer from the solution into the membrane–solution interface, if its concentration is lower than in the membrane–solution interface;

2. complexation of the enantiomer with the chiral selector;

3. when the concentration of the enantiomer in the membrane–solution interface is becoming higher than in the solution, the enantiomer is de-complexed from the complex;

4. re-extraction of the enantiomer from the membrane–solution interface follows its de-complexation.

The main process that is responsible for the potential development is the complexation one.

If L is the chiral selector and S and R the enantiomers to be determined, the following reactions take place:

$$L + S \leftrightarrow LS \tag{1}$$

$$L + R \leftrightarrow LR \tag{2}$$

The stability constants of the complexes formed between chiral selector and enantiomers are given by the following equations:

$$K_S = e^{-\frac{\Delta G_S}{RT}} \qquad K_R = e^{-\frac{\Delta G_R}{RT}}$$

where K_S and K_R are the stability constants of the complexes formed between chiral selector and the S- and R-enantiomer, respectively, ΔG_S and ΔG_R are the free energies of the reactions (1) and (2), respectively, $R = 8.31$ J/K mol and T is the temperature measured in Kelvin.

The enantioselectivity of the chiral selector is given by the difference between the free energies of reactions (1) and (2):

$$\Delta(\Delta G) = \Delta G_S - \Delta G_R$$

A direct proportionality between $\log K$ and ΔG is obvious from the relations between the stability constant and the free energy. This means that a difference in the free energies of the reactions will result in a difference in the stability of the complexes formed between chiral selector and the S- and R-enantiomers.

The stability of the complexes is directly correlated to the slope (response) of the potentiometric, enantioselective membrane electrodes [13]. Accordingly, a large difference between the free energies of reactions (1) and (2) will give a large difference between the slopes when the S- and R-enantiomers are assayed. The enantioselectivity of the measurements will be given by the difference between the two free energies. Also, the slope is the criterion for molecular recognition of the enantiomer, when the electrode is enantioselective. The minimum value admissible for a $1:n$ stoichiometry between the enantiomer and chiral selector is $50/n$ mV/decade of concentration.

2.1.5   Design of Electrochemical Enantioselective Sensors

The design of sensors is a very important tool for enantioselective analysis because its reliability influences the reliability of the analytical information. To be validated, a sensor must have reliable response characteristics. The general validation criteria, given in a previous paper [14], can be re-written and are also valuable for the electrochemical, enantioselective sensors used for enantioselective high-throughput screening of drugs. These criteria are as follows: (1) the best chiral selector must be selected, or the enzyme that favors only the reaction of one enantiomer must be selected; (2) the best matrix for the electroactive material must be used; (3) for in vivo screening analysis, the biocompatibility of materials must be correlated with the response characteristics of the sensors; (4) for the potentiometric, enantioselective membrane electrodes, the minimum slope accepted is 50 mV/decade of concentration, with a low limit of detection, minimum two concentration decades for working concentration range, and maximum selectivity over byproducts, and compression and degradation compounds; (5) high enantioselectivity; (6) a minimum of 99.00% for the recovery test of the enantiomer of interest when its concentration is up to 99.99 times lower than that of the other enantiomer. Practically, the best chosen of the chiral selector and matrix will be able to satisfy reliably points 4 and 5 of the validation criteria.

There are two possibilities to construct enantioselective sensors: (1) by selecting a molecule with a special architecture, or (2) by utilizing a catalyst – enzyme – which favors the reaction of only one enantiomer. Both methods have the same principle: to find a lock for the key, where the lock is the molecule with special architecture or the enzyme, and the key is the enantiomer.

2.1.5.1 Design of Potentiometric, Enantioselective Membrane Electrodes

For the selection of the best chiral selector in enantioselective analysis, combined multivariate regression and neural networks are proposed [15]. The most commonly utilized chiral selectors for potentiometric, enantioselective membrane electrode (PEME) construction include crown ethers [16–19], cyclodextrins [5, 20], maltodextrins [21], and quinine and quinidine derivatives (unpublished data). The response characteristics of these sensors and their enantioselectivity are correlated with the type of matrix used for sensor construction.

Two main types of matrices are described for the design of the sensors: a PVC-based matrix [19, 20] and a carbon paste-based matrix [5]. A special design was adopted for the construction of imprinted polymer-based sensors [22]. The most reproducible design proved to be that based on carbon paste matrix. The nonreproducibility of PVC-based matrices is due to nonuniformity and nonreproducibility of the repartition of the electroactive material in the matrix.

Carbon Paste-Based Potentiometric, Enantioselective Membrane Electrodes

Paraffin oil and graphite powder were mixed in a ratio of 1 : 4 (w/w) followed by the addition of a solution of the chiral selector (10^{-3} mol/L) (100 µL chiral selector solution to 100 mg carbon paste). The graphite–paraffin oil paste was filled into a plastic pipet tip leaving 3–4 mm empty in the top to be filled with the carbon paste that contained the chiral selector. The diameter of the potentiometric, enantioselective membrane sensor was 3 mm. Electric contact was made by inserting a silver wire in the carbon paste [5]. Before each use, the surface of the electrode was wetted with deionized water and then polished with an alumina paper (30144-001 polishing strips, Orion). When not in use, the electrode was immersed in a 10^{-3} mol/L *S*- or *R*-enantiomer solution.

PVC-Based Potentiometric, Enantioselective Membrane Electrodes

Electroactive membranes were prepared incorporating 1.2% chiral selector, 65.6% *o*-nitrophenyl octyl ether (oNPOE) or dioctyl sebacate (DOS) (plastifier), 0.4% tetrakis{[3,5-bis(trifluoromethyl)phenyl]borate} (TKB) in a tetrahydrofuran solution containing the poly(vinyl chloride) (PVC) (32.4% from the membrane composition) [19, 20]. A 7 mm diameter circle was cut from the membrane and inserted into an electrode body along with an inner solution containing the enantiomer to be determined at a certain concentration (usually 10^{-3} mol/L).

2.1.5.2 Design of Biosensors

Two main types of matrices are proposed for enantioselective biosensor design: a polymeric matrix [23, 24], and a matrix based on carbon paste [6, 25, 26]. The most reliable membrane design is that based on carbon paste.

Carbon Paste-Based Amperometric Biosensor

Graphite powder was heated at 700 °C for 15 s in a muffle furnace and cooled to ambient temperature in a desiccator. To 100 mg of activated graphite powder, 100 µL of enzyme solution (1 mg enzyme/mL, made in a certain buffer with a certain pH) were added. This mixture was allowed to react at 4 °C for 2 h before drying under vacuum for 4.5 h to remove water. An aliquot of 40 µL of Nujol (or paraffin) oil per 100 mg of graphite powder was added to dry enzyme-modified graphite to prepare the paste. Plain graphite–Nujol paste was filled into a plastic syringe holder (1.0 mL syringe, ONCE/ASIK, Denmark), leaving about 3–4 mm empty in the top to be filled with chemically modified carbon paste that contains the enzyme. Electric contact was made by inserting a silver wire in the carbon paste. The electrode tips were gently rubbed on fine paper to produce a flat surface. When not in use, the biosensor was stored in a dry state at 4 °C.

Polymer-Based Amperometric Biosensor

Immunodyne membranes (120 µm thickness, 0.3 µm diameter of pores) supplied in a preactivated form by Pall (Glen Cove, NY, USA) were utilized for enzyme immobilization. Each side of the membrane was wetted with 10 µL of enzyme solution at a concentration of 50 mg/mL in a certain buffer. The coupling reaction lasted for 2 min, and then the enzyme membranes, as disks of 8 mm diameter, were washed in 1 mol/L potassium chloride solution for 10 min. They were stored in the buffer at 4 °C.

A design for an enzyme field effect transistor is reported through the immobilization of the enzyme on a pH-selective field effect transistor (pHSFET) [27].

Co-cross-linking with glutaraldehyde, using human serum albumin (HSA) as co-immobilizer, was applied to immobilize the enzyme. The suspended enzyme (in buffer) and lipase (in buffer) were mixed with HSA solution, and with 1 µL of glutardialdehyde. The mixture was immediately applied to the pHSFET. Within 30–90 s a membrane formed; the enzyme was thus immobilized. The enzyme field effect transistors (EnFETs) produced in this manner were washed with distilled water and were immediately ready for use. They were stored in buffer solution at 6 °C.

2.1.5.3   Design of Immunosensors

Antibody proved the capacity of recognition of the chiral center [28]. It follows that the immunosensors can also be used in the enantioselective high-throughput screening of drugs. The design of immunosensors is based on the carbon paste matrix. The enzyme is replaced with the antibody. No preliminary thermal treatment is required for graphite powder [29].

2.1.6   Applications

2.1.6.1   Application of Potentiometric, Enantioselective Membrane Electrodes in Enantioanalysis

Amino Acids and Their Derivatives

L-Proline
A potentiometric, enantioselective membrane electrode based on impregnation of 2-hydroxy-3-trimethylammoniopropyl-β-cyclodextrin (as chloride salt) solution in a carbon paste was proposed for the assay of L-proline [30]. The linear concentration range for the assay of L-proline is between 5.0×10^{-5} and 1.5×10^{-1} mol/L, with a detection limit of 1.0×10^{-5} mol/L. The average recovery is 99.90% with a relative standard deviation (RSD) of 0.12% ($n = 10$). The electrode is enantioselective over D-proline.

L- and D-Dinitrobenzoylleucine (DNB-Leu)
Quinine, quinidine, *tert*-butylcarbamylated quinine and *tert*-butylcarbamylated quinidine were proposed as new chiral selectors in the construction of potentiometric, enantioselective membrane electrodes (unpublished data). A matrix based on carbon paste was used for the design of the electrodes. Electrodes had near-

Nernstian values for the concentrations of low order of magnitude, with low limits of detection ($< 10^{-7}$ mol/L).

Quinine and quinidine proved to be enantioselective for the (S)-DNB-Leu, while *tert*-butylcarbamylated quinine was enantioselective for (R)-DNB-Leu. Accordingly, these electrodes can be used for the enantioselective analysis of one of the enantiomers of dinitrobenzoylleucine. The average recovery of these enantiomers is $> 98.30\%$ with an RSD $< 0.7\%$.

The enantioselectivity of the *tert*-butylcarbamylated quinidine-based electrode is not so good for most of the ratio values between the enantiomers of DNB-Leu, and it cannot be used for the assay of one of the enantiomers unless a semi-quantitative analysis is performed before its utilization.

Angiotensin-Converting Enzyme Inhibitors

(S)-Captopril

(S)-Captopril was the first angiotensin-converting enzyme inhibitor tested with this type of electrochemical sensor [31]. The potentiometric, enantioselective membrane electrode based on impregnation of 2-hydroxy-3-trimethylammonio-propyl-β-cyclodextrin (as chloride salt) solution in a carbon paste can be successfully used for (S)-captopril assay in the 10^{-6}–10^{-2} mol/L concentration range (pH range 3–6.5), with a near-Nernstian slope of 57.70 mV/decade of concentration, a low limit of detection of 2×10^{-7} mol/L, and an average recovery of 99.99% (RSD = 0.05%). Using this sensor, (S)-captopril can also be reliably assayed in pharmaceutical formulations (25 mg (S)-captopril/tablet) with an average recovery of 99.69% (RSD=0.39%). The enantiopurity was tested over (R)-captopril and D-proline; the 10^{-4} order of magnitude obtained for potentiometric selectivity coefficients proved its enantioselectivity.

The response characteristics of the sensor were also tested for (R)-captopril. A non-Nernstian slope of 27.30 mV/decade of concentration and a detection limit of 10^{-6} order of magnitude were obtained.

Maltodextrins (dextrose equivalent (DE) 4–7, 13–17, and 16.5–19.5) have been proposed as novel chiral selectors for the construction of potentiometric, enantioselective membrane electrodes for (S)-captopril assay [21]. The potentiometric, enantioselective membrane electrodes can be used reliably for the assay of (S)-captopril as raw material and from pharmaceutical formulations such as Novocaptopril tablets, using direct potentiometry. The best response was obtained when the maltodextrin with higher DE was used for the electrode's construction. The best enantioselectivity and stability in time were achieved for the lower DE maltodextrin. L-Proline were found to be the main interferent for all proposed electrodes. The surface of the electrodes can be regenerated by simply polishing, obtaining a fresh surface ready to be used in a new assay.

(S)-Cilazapril

For (S)-cilazapril assay [32] the potentiometric, enantioselective membrane electrode, based on impregnation of 2-hydroxy-3-trimethylammoniopropyl-β-cyclodextrin (as chloride salt) solution in a carbon paste, gave a near-Nernstian slope (56.85 mV/decade of concentration) for the 10^{-5}–10^{-2} mol/L concentration range (pH between 3.0 and 5.5), with a limit of detection of 5.0×10^{-6} mol/L. The enantioselectivity of sensor was checked over D-proline; a 10^{-4} order of magnitude potentiometric selectivity coefficient proved the sensors's enantioselectivity. The average recovery of (S)-cilazapril was 99.36% with an RSD of 0.27%.

(S)-Enalapril

(S)-Enalapril assay can be achieved using the potentiometric electrode based on impregnation of 2-hydroxy-3-trimethylammoniopropyl-β-cyclodextrin (as chloride salt) solution in a carbon paste, in the 3.6×10^{-5}–6.4×10^{-2} mol/L (pH between 3.0 and 6.0) concentration range with a detection limit of 1.0×10^{-5} mol/L [33]. The slope is near-Nernstian: 55.00 mV/decade of concentration. The average recovery of (S)-enalapril raw material was 99.96% (RSD=0.098%). The potentiometric selectivity coefficient over D-proline (6.5×10^{-4}) proved the sensor's enantioselectivity. (S)-Enalapril was determined in Renitec tablets with an average recovery of 99.59% (RSD=0.20%).

(S)-Pentopril

Enantioselective determination of (S)-pentopril using a potentiometric, enantioselective membrane electrode, based on impregnation of 2-hydroxy-3-trimethylammoniopropyl-β-cyclodextrin (as chloride salt) solution in a carbon paste can be performed in the 10^{-6}–10^{-2} mol/L (pH=3–5.5) concentration range [32]. The limit of detection was 7.58×10^{-5} mol/L. The slope was near-Nernstian: 58.16 mV/decade of concentration. The potentiometric selectivity coefficient over D-proline was 6.9×10^{-4}. The average recovery of (S)-pentopril was 99.79% (RSD=0.17%).

(S)-Perindopril

The potentiometric, enantioselective membrane electrode, based on impregnation of 2-hydroxy-3-trimethylammoniopropyl-β-cyclodextrin (as chloride salt) solution in a carbon paste, can be used reliably for (S)-perindopril assay in the 10^{-5}–10^{-2} mol/L (pH=2.35–6) concentration range [34]. The detection limit was 5×10^{-6} mol/L and the slope was near-Nernstian: 54.23 mV/decade of concentration. The selectivity was checked over (R)-perindopril and D-proline. The 10^{-4} order of magnitude obtained for the potentiometric selectivity coefficients proved the enantioselectivity of the sensor. For (R)-perindopril, the potentiometric, enantioselective membrane electrode had the following response characteristics: slope non-Nernstian (38.00 mV/decade of concentration), linear range between 10^{-4} and 10^{-2} mol/L, and detection limit 10^{-6} mol/L order of magnitude. (*S*)-Perindopril can be determined with an average recovery of 99.58% (RSD=0.33%).

(S)-Ramipril

(*S*)-Ramipril can be determined as raw material and in its pharmaceutical formulations using the potentiometric, enantioselective membrane electrode, based on impregnation of 2-hydroxy-3-trimethylammoniopropyl-β-cyclodextrin (as chloride salt) solution in a carbon paste, in the 1.8×10^{-5}–2.3×10^{-1} mol/L (pH between 2.5 and 6.0) concentration range with an average recovery of 99.94% (RSD=0.030%) and 98.98% (RSD=0.67%), respectively [35]. The detection limit was 10^{-5} mol/L order of magnitude. The slope was near-Nernstian: 52.00 mV/decade of concentration. Enantioselectivity was proved over D-proline, a 10^{-4} order of magnitude being obtained for the potentiometric selectivity coefficient.

(S)-Trandolapril

The potentiometric, enantioselective membrane electrode, based on impregnation of 2-hydroxy-3-trimethylammoniopropyl-β-cyclodextrin (as chloride salt) solution in a carbon paste, can be used reliably for (*S*)-trandolapril assay with an average recovery of 99.77% (RSD=0.22%) [32]. The linear concentration range was 10^{-4} – 10^{-2} mol/L in the pH range 2.5–5.5. The detection limit was 10^{-5} mol/L order of magnitude. The slope was near-Nernstian: 52.45 mV/decade. The sensor enantioselectivity was determined over D-proline, a 10^{-4} order of magnitude being obtained for the potentiometric selectivity coefficient.

2.1.6.2   Application of Amperometric Immunosensors in Enantioanalysis

An immunosensor based on immobilization of anti-L-T_3 on a carbon paste matrix was proposed for the assay of L-T_3 [36]. Its enantioselectivity versus D-T_3 is high. The main advantages of using such a sensor in enantioselective screening analysis are the highest sensitivity, selectivity, and enantioselectivity and also precision and accuracy that can be obtained.

2.1.6.3   Application of Amperometric Biosensors in Enantioanalysis

L- and D-Amino Acids

For the enantioselective assay of L- and D-amino acids, the corresponding L- and D-amino acid oxidase is utilized in the design of the biosensors. Two types of design have been proposed for the amperometric enantioselective biosensors: carbon paste-based biosensors [37, 38] and screen-printed amperometric biosensors [39].

To optimize the response of the amperometric biosensor, Rivas and Maestroni [37] proposed that the carbon paste should contain 2.3% Ir and 8.5% enzyme. In these conditions, the calibration graphs were linear from 0.1 to 1.5 mmol/L amino acid.

The biosensors designed by screen-printed technology consisted of a rhodinized carbon/hydroxyethylcellulose/polyethylenimine electrode containing immobilized L- or D-amino acid oxidase as the working electrode. All the measurements with these biosensors were made at +400 mV (vs Ag/AgCl) [39]. The calibration graphs were linear up to 1 mmol/L amino acid with detection limits between 0.15 and 0.47 mmol/L. The lifetime of the biosensor was 56 days.

Angiotensin-Converting Enzyme Inhibitors

(S)-Cilazapril

An amperometric biosensor based on L-amino acid oxidase was proposed for the assay of (S)-cilazapril [40]. The enzyme was physically immobilized in a carbon paste. The linear concentration range for the determination of (S)-cilazapril was between 100 and 0.001 µmol/L, with a detection limit of 5 pmol/L. The RSD values of <0.2%, and also its high enantioselectivity over (R)-cilazapril, make the biosensor suitable for the enantiopurity test of (S)-cilazapril raw material and its pharmaceutical formulations.

(S)-Enalapril

An amperometric biosensor based on physical immobilization of L-amino acid oxidase in a carbon paste was proposed for the assay of (S)-enalapril [41]. The linear concentration range was between 0.4 and 120 µmol/L with a detection limit of 163 nmol/L. The working pH range was between 6.8 and 7.4. The amperometric biosensor was enantioselective over the R-enantiomer and also D-proline. The RSD calculated for the enantiopurity tests of (S)-enalapril raw material and of its pharmaceutical formulations was <1% ($n=10$).

(S)-Pentopril

An amperometric biosensor based on the physical immobilization of L-amino acid oxidase was proposed for the enantioselective assay of (S)-pentopril [40]. The linear concentration range (0.08–50 µmol/L), the limit of detection (5 µmol/L), and the high enantioselectivity over the R-enantiomer and D-proline made the amperometric biosensor suitable for the enantiopurity test of S-pentopril raw material and its pharmaceutical formulations. For all the enantiopurity tests, the RSD values did not exceed 0.2% ($n=10$).

(S)-Perindopril

Because (S)-perindopril is the enantiomer that is responsible for the angiotensin-converting enzyme inhibition activity, it is necessary to develop a reliable method for its enantioselective assay. An amperometric biosensor based on L-amino acid oxidase physically immobilized on a carbon paste matrix was proposed in this regard [42].

The working range of the amperometric biosensor was between 20 pmol/L and 10 μmol/L in the pH range between 7.0 and 7.4, with a detection limit of 2 pmol/L. The lifetime of the biosensor was 3 weeks. The amperometric biosensor can be used for enantioselective analysis of (S)-perindopril raw material and its pharmaceutical formulations with an RSD <1% ($n=10$).

(R)-Perindopril

An amperometric biosensor based on D-amino acid oxidase was described for the assay of (R)-perindopril [9]. (R)-Perindopril can be determined in the 400–20 nmol/L concentration range, with a limit of detection of 10 nmol/L.

Its enantioselectivity was checked over S-perindopril and L- and D-proline. The amperometric biosensor showed high enantioselectivity over (S)-perindopril and L-proline. The RSD values were <0.04% when the electrode was used for the assay of (R)-perindopril.

(S)-Ramipril

An amperometric biosensor based on L-amino acid oxidase was developed and proved reliable for the analysis of the (S)-ramipril [41]. This biosensor is based on the physical immobilization of L-amino acid oxidase in a carbon paste.

The linear concentration range of the amperometric biosensor was between 100 and 0.2 μmol/L with a limit of detection of 107 nmol/L. The working pH range was between 6.2 and 7.0. The RSD of <1% ($n=10$) assured the suitability of the amperometric biosensor for the enantioselective analysis of (S)-ramipril in raw material and its pharmaceutical preparations.

(S)-Trandolapril

An amperometric biosensor based on the physical immobilization of L-amino acid oxidase was recommended for the enantioselective assay of (S)-trandolapril [40]. This amperometric biosensor had a linear concentration range between 10 and 0.02 μmol/L with a detection limit of 15 μmol/L. The RSD values of <0.2% ($n=10$) and its high enantioselectivity made the amperometric biosensor suitable for the enantioselective analysis of (S)-trandolapril.

2.1.6.4 L-Ascorbate

An enzyme-less amperometric biosensor based on poly-L-histidine–copper complex as an alternative biocatalyst was proposed for the enantioselective assay of L-ascorbate [43]. The biosensor was designed by entrapping poly-L-histidine in polyacrylamide gel and subsequent immersion of the gel in $CuCl_2$ solution for 10 min. The resulting membrane was then placed over the gas-permeable membrane of a Clark-type O_2 electrode. The calibration graph was linear between 3 and 300 μmol/L with an RSD of <3% ($n = 10$) at the 100 μmol/L level.

2.1.6.5 D-Fructose

A method for the determination of D-fructose was based on the utilization of an amperometric biosensor with a PVC binding matrix [44]. The enzyme used was D-fructose dehydrogenase. The amperometric signals were fast, reproducible, and linearly proportional to D-fructose concentration from 10 to 0.05 mmol/L.

2.1.6.6 L-Glutamate

L-Glutamate was determined using a biosensor based on carbon paste wax electrode with thermophilic L-glutamate dehydrogenase, NADP, a polymeric toluidine blue, and hexaamineruthenium(III) trichloride [45]. The calibration graph was linear up to 40 mmol/L L-glutamate, with a detection limit of 0.3 mmol/L and an RSD of 7.6% ($n = 10$).

2.1.6.7 L-Lactate

An amperometric biosensor based on the immobilization of bacterial cells of *Alcaligene eutrophud* KTO_2 on an oxygen electrode was proposed for the assay of L-lactate [46]. The biosensor can be used for the enantioselective assay of L-lactate even in a mixture containing acetate and succinate.

An alternative to this amperometric biosensor is the utilization of an amperometric biosensor based on glucose oxidase and on violuric acid as electron transfer mediator [47]. A graphite rod was used as a matrix in the amperometric biosensor design. The amperometric biosensor can be used for the assay of L-lactate in the 12–2 mmol/L concentration range.

2.1.6.8   D-Lactic Acid

A microsensor based on silicon films with DNA-dependent dehydrogenase was proposed for the assay of D-lactic acid [48]. The response of the electrode was linear up to 1.1 mmol/L. The lifetime of the electrode was 3 weeks.

2.1.6.9   L-Lysine

A highly enantioselective and fast amperometric biosensor was based on the immobilization of L-lysine oxidase and rhodium on a ruthenium-immobilized poly-1,2-diaminobenzene-coated vitreous carbon electrode [49]. The limit of detection was low. No other L- or D-amino acids responded to the sensor. The reliability of this amperometric biosensor was high.

2.1.6.10   D-Lysine

A benzoquinone-mediated amperometric biosensor based on D-amino acid oxidase was proposed for the assay of D-lysine [50]. The calibration graphs of the sensor response at 150 mV were linear in the concentration range between 10 and 1 mmol/L of D-lysine.

2.1.6.11   L-Malate

A biosensor for the determination of L-malate in wine was designed by depositing L-lactate dehydrogenase and diaphorase on the surface of solid composite transducer composed of 2-hexadecanone, graphite, and NAD^+ [51]. A linear response was obtained up to 1.1 mmol/L of L-malate with a detection limit of 10 μmol/L.

2.1.6.12   L-Tryptophan

A tryptophan–2-monooxygenase-based amperometric biosensor was proposed for the assay of L-tryptophan [52]. A linear response was recorded for concentrations between 2 and 0.1 mmol/L. The amperometric biosensor can be used for the determination of the concentration of L-tryptophan in nutritional broths.

2.1.7 References

[1] Aboul-Enein, H.Y., Wainer, I.W., *The Impact of Stereochemistry on Drug Development and Use*; New York: Wiley, 1997.

[2] Stefan, R.I., van Staden, J.F., Aboul-Enein, H.Y., *Crystal Eng.* **4** (2001) 113–118.

[3] Stefan, R.I., van Staden, J.F., Aboul-Enein, H.Y., *Combinat. Chem. High Throughput Screening* **6** (2000) 445–454.

[4] Stefan, R.I., van Staden, J.F., Aboul-Enein, H.Y., *Electroanalysis* **11** (1999) 1233–1235.

[5] Aboul-Enein, H.Y., Stefan, R.I., van Staden, J.F., *Anal. Lett.* **32** (1999) 623–632.

[6] Stefan, R.I., Radu, G.L., Aboul-Enein, H.Y., Baiulescu, G.E., *Curr. Trends Anal. Chem.* **1** (1998) 135–138.

[7] Aboul-Enein, H.Y., Stefan, R.I., *Crit. Rev. Anal. Chem.* **28** (1998) 259–266.

[8] Stefan, R.I., van Staden, J.F., Aboul-Enein, H.Y., *Sens. Actuators B* **54** (1999) 261–265.

[9] van Staden, J.F., Stefan, R.I., Aboul-Enein, H.Y., *Fresenius' J. Anal. Chem.* **367** (2000) 178–180.

[10] van Staden, J.F., Stefan, R.I., Aboul-Enein, H.Y., *Anal. Chim. Acta* **411** (2000) 51–56.

[11] Stefan, R.I., van Staden, J.F., Aboul-Enein, H.Y., *Talanta* **51** (2000) 969–975.

[12] Stefan, R.I., van Staden, J.F., Aboul-Enein, H.Y., *Biosens. Bioelectron.* **15** (2000) 1–5.

[13] Stefan, R.I., Aboul-Enein, H.Y., *Instrum. Sci. Technol.* **27** (1999) 105–110.

[14] Stefan, R.I., Aboul-Enein, H.Y., *Accred. Qual. Assur.* **3** (1998) 194–196.

[15] Booth, T.D., Azzaoui, K., Wainer, I.W., *Anal. Chem.* **69** (1997) 3879–3883.

[16] Yasaka, Y., Yamamoto, T., Kimura, K., Shono, T., *Chem. Lett.* **20** (1980) 769–772.

[17] Shinbo, T., Yamaguchi, T., Sakaki, K., Yanagishita, H., Kitamoto, D., Sugiura, M., *Chem. Express* **7** (1992) 781–784.

[18] Bussmann, W., Lehn, J.M., Oesch, U., Plumere, P., Simon, W., *Helv. Chim. Acta* **64** (1981) 657–661.

[19] Horvath, V., Takacs, T., Horvai, G., Huszthy, P., Bradshaw, J.S., Izatt, R.M., *Anal. Lett.* **30** (1997) 1591–1609.

[20] Kataky, R., Parker, D., Kelly, P.M., *Scand. J. Clin. Lab. Invest.* **55** (1995) 409–419.

[21] Stefan, R.I., van Staden, J.F., Aboul-Enein, H.Y., *Fresenius' J. Anal. Chem.* **370** (2001) 33–37.

[22] Zhan, S.Z., Dai, Q., Yuan, C.W., Lu, Z.H., Haeussling, L., *Anal. Lett.* **32** (1999) 677–687.

[23] Radu, G.L., Coulet, P.R., *Rev. Roum. Biochim.* **29** (1992) 239–244.

[24] Radu, G.L., Coulet, P.R., *Analusis* **21** (1993) 101–105.

[25] Johansson, E., Marko-Varga, G., Gorton, L., *J. Biomater. Appl.* **8** (1993) 146–173.

[26] Kacaniklic, V., Johansson, K., Marko-Varga, G., Gorton, L., Jonsson-Pettersson, G., Csoregi, E., *Electroanalysis* **6** (1994) 381–390.

[27] Kullick, T., Ulber, R., Meyer, H.H., Schepper, T., Schugerl, K., *Anal. Chim. Acta* **293** (1994) 271–276.

[28] Hofstetter, O., Hofstetter, H., Schurig, V., Wilchek, M., Green, B.S., *J. Am. Chem. Soc.* **120** (1998) 3251–3252.

[29] Aboul-Enein, H.Y., Stefan, R.I., Radu, G.L., Baiulescu, G.E., *Anal. Lett.* **32** (1999) 623–632.

[30] Stefan, R.I., van Staden, J.F., Aboul-Enein, H.Y., *Anal. Lett.* **31** (1998) 1784–1794.

[31] Stefan, R.I., van Staden, J.F., Aboul-Enein, H.Y., *Talanta* **48** (1999) 1139–1143.

[32] Stefan, R.I., van Staden, J.F., Aboul-Enein, H.Y., *Electroanalysis* **11** (1999) 192–194.

[33] Aboul-Enein, H.Y., Stefan, R.I., van Staden, J.F., *Analusis* **27** (1999) 53–56.

[34] Stefan, R.I., van Staden, J.F., Aboul-Enein, H.Y., *Chirality* **11** (1999) 631–634.

[35] Stefan, R.I., van Staden, J.F., Baiulescu, G.E., Aboul-Enein, H.Y., *Chem. Anal. (Warsaw)* **44** (1999) 417–422.

[36] Aboul-Enein, H.Y., Stefan, R.I., Radu, G.L., Baiulescu, G.E., *Anal. Lett.* **32** (1999) 447–455.

[37] Rivas, G.A., Maestroni, B., *Anal. Lett.* **30** (1997) 489–501.

[38] Dricks, J.M., Aston, W.J., Davis, G., Turner, A.P.F., *Anal. Chim. Acta* **182** (1986) 103–112.

[39] Sarkar, P., Tothill, I.E., Setford, S.J., Turner, A.P.F., *Analyst* **124** (1999) 865–870.

[40] Aboul-Enein, H.Y., Stefan, R.I., Radu, G.L., *Pharm. Dev. Technol.* **4** (1999) 251–255.

[41] Stefan, R.I., Aboul-Enein, H.Y., Radu, G.L., *Prep. Biochem. Biotechnol.* **28** (1998) 305–312.

[42] Aboul-Enein, H.Y., Stefan, R.I., Radu, G.L., *Prep. Biochem. Biotechnol.* **29** (1999) 55–61.

[43] Hasebe, Y., Akiyama, T., Yagisawa, T., Uchiyama, S., *Talanta* **47** (1998) 1139–1147.

[44] Stred'ansky, M., Pizzariello, A., Stredanska, S., Miertus, S., *Anal. Commun.* **36** (1999) 57–61.

[45] Pasco, N., Jeffries, C., Davies, Q., Downard, A.J., Roddick-Lanzilotta, A.D., Gorton, L., *Biosens. Bioelectron.* **14** (1999) 171–178.

[46] Plegge, V., Slama, M., Sueselbeck, B., Wienke, D., Spener, F., Knoll, M., Zaborosch, C., *Anal. Chem.* **72** (2000) 2937–2942.

[47] Krikstopaitis, K., Kulys, J., *Electrochem. Commun.* **2** (2000) 119–123.

[48] Tap, H., Gros, P., Gue, A.M., *Electroanalysis* **11** (1999) 973–977.

[49] Kelly, S.C., O'Connell, P.J., O'Sullivan, C.K., Guilbault, G.G., *Anal. Chim. Acta* **412** (2000) 111–119.

[50] Murphy, A.S.N., Sharma, J., *Electroanalysis* **11** (1999) 188–191.

[51] Katrlik, J., Pizzariello, A., Mastihuba, V., Svorc, J., Stred'ansky, M., Miertus, S., *Anal. Chim. Acta* **379** (1999) 193–200.

[52] Simonian, A.L., Rainina, E.I., Fitzpatrick, P., Wild, J.R., *Biosens. Bioelectron.* **12** (1997) 363–371

List of Symbols and Abbreviations

Symbol	Designation
ΔG	free energy
K	stability constant
R	gas constant
T	absolute temperature

Abbreviation	Explanation
AAOD	amino acid oxidase
Ab	antibody
Ag	antigen
DE	dextrose equivalent
DNB-Leu	dinitrobenzoylleucine
ELISA	enzyme-linked immunosorbent assay
FIA	flow injection analysis
HSA	human serum albumin
PEME	potentiometric, enantioselective membrane electrode
pHSFET	pH-selective field effect transistor
PVC	poly(vinyl chloride)
RSD	relative standard deviation
SIA	sequential injection analysis

2.2 Electronic Tongues: Sensors, Systems, Applications

A. Legin, A. Rudnitskaya, Y. Vlasov,
St. Petersburg University, St. Petersburg, Russia

Abstract

Multisensor systems for liquid analysis based on chemical sensor arrays and pattern recognition, which are now widely known as 'electronic tongues', represent one of the most rapidly emerging and exciting fields of non-classical analytics during the last decade. This chapter presents an overview of the research and development of electronic tongue systems and describes various sensors, sensor arrays, and their numerous applications. A sound basis for electronic tongues was provided by the extensive development of well-known selective sensors, especially electrochemical, and biological inspirations originating from sensory systems of mammalians. The up-to-date achievements of various scientific groups working in this field are reviewed. The performance of electronic tongues in the tasks of recognition (classification, identification, discrimination) of multicomponent media is considered. A useful option of multicomponent quantitative analysis with the help of electronic tongues is also reported. The correlation between the output of an electronic tongue and human sensory assessments of food flavor made by taste panel opens up an exciting possibility of measuring and quantifying the taste and flavor of foods. Application areas of the electronic tongue systems, including quality control of foodstuffs, clinical, industrial and environmental analysis, are surveyed. Future prospects for research and development of electronic tongues are discussed.

Keywords: Electronic tongue; sensor systems; cross-sensitivity; quantitative and qualitative analysis; food flavor assessment

Contents

2.2.1   Introduction

It may be easier to understand a new emerging idea or methodology by considering it in relation to prior art and other contributing areas and inputs. Before describing electronic tongues themselves, it seems useful to give a brief overview of background knowledge and inspirations.

2.2.1.1   From Discrete Sensors to Sensor Arrays

R&D of analytical instruments was traditionally aimed at obtaining the highest possible selectivity for an analyte. In the field of chemical sensors for solution analysis, these efforts resulted in the development of multiple electrochemical sensors and, in particular, of potentiometric sensors, widely known as ion-selective electrodes (ISEs). Since most currently known and advanced electronic tongues originate genetically from ISEs, they can provide a good illustration of the evolution of chemical sensors and sensor arrays and related ideas.

The first ISE with an oxide glass sensitive membrane for the determination of hydrogen activity in aqueous solutions was suggested in 1907 by Haber and Klemensiewicz [1]. Since then, a significant number of ISE membrane materials, both inorganic and organic, have been implemented. The main types of ISEs include oxide glasses for H^+ and alkali and alkali earth metal cation determination, crystalline materials for determination of halides and heavy metals, liquid or plasticized polymer compositions with ion exchangers or neutral carriers, chalcogenide glasses for heavy metal determination, and membranes with immobilized enzymes for detection of some organic substances. Some important features of ISEs are as follows: the concentration (activity) of ionic forms but not the total content is being measured, an ISE response is linear, and direct potentiometric measurements are often possible. Extensive descriptions of existing ISEs can be found in several books and reviews, eg [2–7]. However, in spite of major efforts by numerous researchers, the pH glass electrode still remains the most selective ISE and the most commonly used. On the other hand, owing to the ease of handling, short analysis time, obvious possibility of automation, an option of miniaturization and low cost, ISEs are attractive tools for the analysis of solutions. In practice, this is particularly true for media where the content of interfering species is sufficiently low. However, the lack of selectivity significantly limits the practical application of many ISEs in the presence of other species in solutions, besides primary ions, and this is a common case in the real-world sample analysis, which is the most important field.

Traditional potentiometry assumes a linear dependence between an ISE output (electrical potential) and logarithm of activity of the primary ion in a solution. The electrode response should obey the Nernst equation, which is commonly used to construct a calibration curve:

$$E = E^\circ + \frac{RT}{z_i F} \ln a_i \tag{1}$$

where E is the potential difference (e.m.f.) of the electrochemical cell comprising an ion-selective and a reference electrode, E° is the standard potential, R is the gas constant, T is the absolute temperature, F is the Faraday constant, z_i is the electrical charge of the primary ion, and a_i is the activity of the primary ion. The term $RT/z_i F$ is known as the response slope S, that is, the sensitivity of an ISE.

The influence of interfering ions on the ISE response is described by the Nikolsky equation:

$$E = E^\circ + \frac{RT}{z_i F} \ln \left[a_i + \sum_j K_{ij}(a_j)^{z_i/z_j} \right] \tag{2}$$

where K_{ij} is the selectivity coefficient of the ISE to the primary ion i in the presence of an interfering ion j and z_i and z_j are the charges of the primary and interfering ions, respectively.

If an ISE is not highly selective, the value of its output (potential) will be determined by the simultaneous presence and the ratio of the contents of several ions or other species. The terms a_i and $\sum_j K_{ij}(a_j)^{z_i/z_j}$ in Equation (2) may appear comparable and the electrode response becomes nonlinear. Although it is still possible to deal with a nonlinear calibration curve, more than one electrode is needed to find correctly the parameters of Equation (2) for multiple analytes. Evidently, the number of electrodes should not be less than the number of analytes according to simple mathematical considerations.

In the mid-1980s, this reasoning led to the idea of applying an electrode array instead of a discrete ISE with the aim of improving the insufficient selectivity of the ISEs in the presence of interfering ions. This approach assumes that the behavior of each electrode of the array in multicomponent solutions can still be described by the Nikolsky equation. Thus, the system of Nikolsky equations should be solved to find ISE parameters such as standard potentials, selectivity coefficients, and/or slopes. The parameters found can be used subsequently for the prediction of the concentration of multiple analytes. Numerous methods can be applied for calculating the parameters of Equation (2), including different regression techniques and even artificial neural networks (ANNs).

The first work dealing with the application of ISE arrays for multicomponent analysis was by Otto and Thomas in 1985 [8]. A sensor array comprising eight sensors was used for the simultaneous determination of sodium, potassium, calcium, and magnesium at the concentration levels typical for biological liquids. The main problem in this case is the insufficient selectivity of Mg- and Na-selective electrodes in the presence of calcium and potassium, respectively. Multiple linear and partial least square (PLS) regressions were used for fitting two parameters of the Nikolsky equation, the standard potential and the selectivity coefficient, the slope values being preliminarily determined in the individual solutions of analytes. The array gave a certain advantage in comparison with discrete ISEs. The best results were obtained using PLS regression.

Kowalski and co-workers [9, 10] used a sensor array comprising five sparingly selective electrodes for the determination of Na^+ and K^+ in binary solutions. A nonlinear regression based on a simplex algorithm and multiple linear regression were used to determine the parameters of the Nikolsky equation including the slope values. The calibration of a sensor array in [9] was performed using a projection pursuit regression, which is a nonparametric multivariate technique, that requires no a priori information about the functional form of the sensor response. The best results were achieved using the nonlinear regression, the determination errors being 0.4% for Na^+ and 5.3% for K^+.

A combination of three highly selective electrodes (sodium, potassium, and calcium) and one sparingly selective sensor with multiple ionophores was employed for the determination of sodium, potassium, and calcium in tertiary mixtures [11]. It was found that incorporation of the fourth sensor (a sparingly selective one) in the array of ISEs decreased the error of determination to 2.8% compared with 4.5% with only three highly selective sensors. The same sensor array was applied in the flow injection set-up for Na^+, K^+, and Ca^+ determination in mineral water and human blood plasma samples [12, 13].

The possibility of using ANNs for processing of sensor array signals was also studied [14]. The measurements were carried out using a flow injection system with the same three ISEs as described in [13]. The samples contained sodium, potassium, and calcium at different concentration levels. The feed-forward neural network was trained to detect and identify ions in a sample. The network parameters and training algorithm were varied to investigate their influence on the neural net performance. Electrode responses were distorted in different ways, including noise addition and baseline shift, with the aim of investigating the ability of the network to recognize previously unseen samples.

Van der Linden and co-workers [15, 16] applied ANNs for the calibration of sensor arrays. A back-propagation neural net was used to obtain a calibration model, and a recurrent one was applied for the determination of Nikolsky equation parameters. Ion-selective sensor arrays were employed for the determination of calcium and copper and of potassium, calcium, nitrate and chloride simultaneously. Calcium- and copper-selective electrodes together with a pH electrode were used in the former case. An average error of about 8% for Ca^{2+} and Cu^{2+} determination was reported. The second array included corresponding ISEs and also a pH electrode. The average error was found to be 6% and the maximum error was about 20%.

These studies illustrate the first attempts to use an array of potentiometric sensors instead of discrete sensors for quantitative ion determinations. The starting point was the idea of using the Nikolsky equation instead of the Nernst equation to take into account interferences from other ions. This resulted in applying sparingly selective electrodes that are more useful in the arrays than more selective electrodes and increasing interest in nonparametric data processing techniques, which do not require information about the functional form of the sensor signal/analyte activity dependence.

2.2.1.2 Biological Inspiration

Another approach that encouraged the development of chemical multisensor systems was an attempt to mimic the organization and performance of biological sensory systems, particularly the olfaction of mammalians (the sense of smell) [17]. These principles were partly realized in systems for gas analysis termed 'electronic noses' and later in liquid analyzers termed 'electronic tongues'. Olfaction was recognized long ago as the most effective sensing system owing to its high sensitivity and discrimination ability. The sense of smell is capable of distinguishing thousands of different volatile molecules, including some very similar ones such as stereoisomers. The perception threshold of humans for some odorants can be as low as a few parts per trillion and it is even lower in animals [18]. Evidently, all odorant substances are volatile but volatility (vapor pressure) and odor intensity are not proportional and some compounds with very low vapor pressure can be powerful odorants (eg, musk) and vice versa. The relationship between a compound structure and the odor that it elicits is still unclear and compounds of very different chemical compositions may have similar odors. However, a relation between fat solubility and odor intensity was postulated: the strongest odorants are both water and fat soluble [19]. An impressive performance of the olfactory system is achieved owing to a wide set of nonspecific or cross-sensitive receptors and processing of their signals in the neural system and in the brain. The detection of odor is performed by olfactory receptor neurons situated in mucus layer in the nostrils, where odorant molecules react with odorant-binding proteins – the 'sensing layer' of receptors. Since receptors are not selective, many of them respond to a given odor. This reaction results in an activity pattern which is transferred to the olfactory bulb, where primary signal processing is performed, and then to the higher level brain region for identification and recognition.

The sense of taste in mammalians is organized similarly to olfaction. Taste is perceived by nonspecific taste buds, situated on the papillae of the tongue. Conventionally, the overall taste is correlated with a combination of four basic tastes: sweetness, sourness, bitterness, and saltiness. Sometimes another elementary taste characteristics are used, such as umami. Umami was firstly introduced by Japanese researchers and it is described as a delicious taste perceived in meat, cheeses, and mushrooms [20]. Since taste and odor are often perceived simultaneously, the term 'flavor' is widely used to describe their combination, especially when speaking about food.

The relationship between taste (flavor) and chemical composition is often not known precisely, especially for sweet substances. Another interesting and highly controversial issue is the interaction between different tastes [19]. In most cases a desensitizing effect or threshold increase takes place when two substances eliciting different tastes are present simultaneously. A further effect is the sensitivity threshold decrease when substances present at nonperceptible concentrations can be felt if a contrasting taste substance is applied to the tongue. Perception thresholds of the human tongue to most of taste substances are much higher than those for olfaction with exception of alkaloids, such as quinine. However, differential taste and odor thresholds are comparable [19]. Thus, the mammalian sense of taste function is similar to olfaction but is less developed, possibly because it is less related to the survival of living beings.

The spectacular capabilities of biological sensory systems inspired scientists to implement their organization principles in artificial sensory devices. The latter were first intended for gas analysis and odor recognition. According to [17], the first attempt to develop an odor detection system dates back to the early 1960s [21]. The history of intelligent multisensor systems for gas analysis started in 1982 with the work of Persaud and Dodd [22]. Since that time, many different groups have tried to add to the development and application of such devices, which were named 'electronic noses' [17, 23]. 'Noses' usually provide for the qualitative recognition of gas mixtures and/or the identification of certain individual gases, eg, from leakages of chemicals.

The same principles were applied to the development of multisensor systems for liquid analysis – 'electronic tongues'. Although an electronic tongue like a biological one works in liquid media, the sensitivity and detection threshold of artificial 'tongues' could be much better, which makes their performance more similar to olfactory system. Furthermore, many, if not all, volatiles originate from either liquid or solid media, eg, in foodstuffs. In fact, the electronic tongue can be thought of as analogous to both olfaction and taste sense and it can be used for the detection of all types of dissolved compounds, including volatile compounds that give odors after evaporation.

A common feature of all electronic nose and electronic tongue systems is the combination of an array of nonspecific sensors together with data processing by pattern recognition methods. Unlike the electronic nose systems, however, the electronic tongues are much more widely employed not only for recognition and classification but also for the quantitative determination of multiple component concentrations. On the other hand, the electronic tongue approach significantly widens the application areas of chemical sensors because of the possibility of performing classification and recognition of complex liquids, which is a very un-typical type of analysis with either ISEs or any other traditional analytical tech-niques.

Summarizing, we can define the electronic tongue as an analytical instrument comprising an array of non-selective chemical sensors with partial specificity to different components in liquids and an appropriate pattern recognition or multi-variate calibration tool, capable of recognizing the quantitative and qualitative composition of simple and complex solutions.

2.2.2   Review of Electronic Tongue Systems

2.2.2.1   Cross-sensitivity

The most important part of an electronic tongue is obviously a sensor array. There is experimental evidence that cross-sensitive or sparingly selective sensors are more useful in multisensor system than more selective sensors [8–16]. The reproducibility of the sensor response is, of course, crucial in all cases. Thus, the sensors to be used in the array should display reproducibility and cross-sensitiv-ity, which is understood as sensitivity to many components of the analyte medi-um simultaneously. The term cross-sensitivity is commonly used for describing sensor properties in the literature devoted to multisensor systems. However, neither a theoretical description of this property nor even its unambiguous and conventional definition is yet available. Closely related terms are also numerous (nonselectivity, cross-reactivity, partial specificity, global selectivity, etc.) and are often used interchangeably. Here we will stick to the term cross-sensitivity since it seems the best approximation to the phenomenon itself.

Cross-sensitivity of potentiometric sensors cannot be treated simply as a reverse value of the selectivity coefficient. The 'classical' selectivity of ISEs was always considered in the framework of a thermodynamic approach, on the basis of certain sensing mechanisms (ion exchange) and for the situation when one primary and one interfering ion are present. Recently, equations describing the general sensor-mixed response were suggested [24]. This approach takes into account the response of a polymer-based potentiometric sensor to any number of ions of different charges. It is based on the phase boundary potential model and assumes that the sensitivity mechanism is still ion exchange. The sensing mechanism, however, may be different or even variable, eg, some sensors would respond both to ionic and to nonionic species in solutions. An adequate theoretical consideration of cross-sensitivity seems not to be possible at the current stage and much more experimental evidence and theoretical considerations of the sensing mechanisms of different materials to different substances, including nonionic ones, are needed. However, an empirical method of sensor cross-sensitivity assessment, which can be used to guide sensor choices in practical applications, was suggested [25, 26].

The first necessary step is the determination of a set of substances, for which cross-sensitivity is to be studied, and a set of sensing materials. In most cases even 'very' nonselective or cross-sensitive sensors would not respond to any ion or substance in solution, but presumably to a certain group of substances. The calculation of cross-sensitivity parameters was based on the sensitivity study of chalcogenide glass electrodes to a set of heavy metals [25, 26]. Later the same parameters were successfully applied to the cross-sensitivity evaluation of other types of membrane materials on different sets of analytes [27]. The experimental measurements used for cross-sensitivity estimation were simply calibrations of given sensors in individual solutions of the chosen set of compounds.

After the application of different fitting procedures and consideration of literature data, the following three parameters were chosen for the description of integral sensor response and cross-sensitivity. As the parameters involved are empirical ones, it is possible to suggest another version or set of them. However, these appeared to be sufficiently representative and successful.

The first is the average sensor response slope S, measured in solutions of the chosen set of substances:

$$S = \frac{1}{n}\sum_i S_i \tag{3}$$

where S_i is sensor response slope in solutions of each individual substance and n is the number of components in the set.

The second value is the average signal-to-noise ratio of a sensor (for all components of the set):

$$K = \frac{1}{n}\sum_i K_i = \frac{1}{n}\sum_i \frac{S_i}{s_i^2} \tag{4}$$

where S_i and s_i are response slope and its standard deviation in solutions of each substance.

The last parameter is termed the 'nonselectivity factor', because it describes the distribution of the sensitivity of a sensor to different components from the chosen set and is calculated as follows:

$$F = \frac{S}{s^2} \tag{5}$$

where S is the average slope (the first parameter) and s is its standard deviation.

The average slope value is the main and the most important characteristic of integral response and, hence, cross-sensitivity of a sensor. The higher its value, the better is the overall sensitivity of the sensor to the substances from the set. An optimal range of the average slope could be estimated in each case on the basis of the following suppositions. Let us consider, eg, a study of sensor sensitivity to a set of divalent ions. In this case, the sensor response slopes are likely to fall in the range from 0 to 29 mV/pX according to the Nernst equation. However, a super-Nernstian response may also be observed. Therefore, the average slope of a sensor close to 30 mV/pX is commonly related to a comparatively uniform distribution of sensitivity to the chosen set of divalent ions. A value of $S > 30$ mV/pX may be a result of a significantly super-Nernstian response to one of the ions. Thus, in this particular case the range of S from 25 to 30 mV/pX should be considered as the optimal one. The sensors displaying $S > 25$ mV/pX display remarkable cross-sensitivity to all ions of the set and can be used for multisensor array analysis.

The average slope value is not the only valid response characteristic. At least two other measures appeared to be useful. To characterize the distribution of sensitivity for different components, the non-selectivity parameter F has been used. $F \ll 0.1$ is typical for highly selective sensors with high sensitivity to the primary ion and very poor sensitivity to the other ions. An increase in F up to 0.1 is evidence for a smoother distribution, but the sensitivity to some components can still be low. A fairly uniform sensitivity distribution to most substances from the set is typical for sensors with $F \geq 1$. Finally, values of $F > 0.5$ characterize a reasonable distribution of sensitivity and significant cross-sensitivity to different species in complex solutions and thus may be considered as optimal for sensors designed for array applications.

The stability parameter is also important because in preliminary experiments a correlation between stability in individual ion solutions and that in complex liquids was found. The average signal-to-noise ratio K is a valuable estimate of sensor stability. The higher is K, the more reproducible is the sensor potential and the more stable is the electrochemical sensor behavior both in individual solutions and mixed liquid media. It was determined experimentally that values of $K > 2$ could be used as a measure of reasonable sensor stability for array applications.

In conclusion, it must be noted that the exact optimal values of cross-sensitivity parameters should be determined in each case individually. In particular, the value of the average slope can vary significantly. The parameters for cross-sensitivity estimation were developed and applied only for potentiometric chemical sensors. However, since no assumptions about the mechanism or theoretical description of sensor response were considered for cross-sensitivity parameter assessments, but only the experimental response value and its standard deviation, the same method could be applied for other types of sensors.

2.2.2.2   Data Processing

The other important aspect of multisensor analysis, apart from the sensor arrays, is the signal processing. In a multicomponent environment the sensor array produces complex signals (patterns), which contain information about different compounds and other features. These signals should be analyzed together to extract valuable analytical information. Various methods of multivariate calibration and pattern recognition are now available and can be used for sensor array data processing. The electronic tongue may be applied in principle to two main tasks: quantitative determination of the content of components and classification (recognition, identification, discrimination). The choice of the data processing technique for a particular case depends on the task to be solved and the structure of the data (nonlinearity, correlations, etc.). A brief overview of some methods, which are most often used in this field, together with their main features, is shown in Table 2.2-1. A more detailed description of data processing methods in sensor analysis is beyond the scope of this chapter. Theoretical discussions of available methods and case studies of different sensor applications can be found in numerous books, manuals and papers [28–30 and references cited therein].

Table 2.2-1. Selected methods of multivariate calibration and pattern recognition used for electronic tongue data processing

Method	Linear	Supervised	Advantage	Drawback
PCA	Yes	No	Easy to interpret	Sensitive to the drift in the data
PLS	Yes	Yes	Statistical description of the results Small calibration data set	
SOM	No	No	2D representation of the data of any dimensionality	Works as black box
BPNN	No	Yes	Easily deals with nonlinear data	

2.2.3   Sensors for Electronic Tongues

It was mentioned in Section 2.2.1 that the first studies on the application of sensor arrays for multicomponent analysis of liquids were performed using potentiometric chemical sensors (ISEs). Potentiometric sensors still remain the most widely used type in electronic tongue systems. However, multisensor systems based on the same principles could also be realized with other types of liquid sensors. In this section, the main types of sensors that have been implemented in electronic tongue systems are discussed.

The first multisensor system for liquid analysis, which could be called an electronic tongue in the proper sense of the term, was a 'taste sensor' (recently referred to by the authors also as the electronic tongue), introduced in 1990 by Toko and co-workers [31, 32] of Kyushu University, Japan. The taste sensor consisted of eight potentiometric sensors with thick-film polymer membranes based on poly(vinyl chloride) (PVC). The membranes contained dioctyl phenylphosphonate (DOPP) as plasticizer and active substances called 'lipids' by the authors, tetrahydrofuran being used as a solvent [33–35]. Membrane compositions are given in Table 2.2-2. These membranes were used for the preparation of potentiometric sensors with a liquid inner filling. Potential values of the sensors were measured versus the conventional Ag/AgCl electrode. Potentiometric measurements were made using an eight-channel scanner connected to the sensors through a high-input impedance amplifier. The manipulations with the sensor array were done by a robot arm, the overall system performance being controlled by computer. A schematic diagram of the taste sensor is shown in Figure 2.2-1.

The device was named a taste sensor because it was claimed to perceive the taste of food in the same manner as by humans. The sensors with PVC membranes containing 'lipids' are supposed to differentiate between tastes instead of detecting each substance selectively. This means that the sensor should respond similarly to substances eliciting similar tastes independently of their chemical structure.

Table 2.2-2. Lipid materials used in the multichannel electrode. Reprinted from Toko, K., Taste sensor, *Sens. Actuators B* **64** (2000) 205–215, with permission from Elsevier Science.

Channel	Lipid (abbreviation)
1	Decyl alcohol (DA)
2	Oleic acid (OA)
3	Dioctyl phosphate (DOP)
4	DOP:TOMA = 9:1
5	DOP:TOMA = 5:5
6	DOP:TOMA = 3:7
7	Trioctylmethylammonium chloride (TOMA)
8	Oleylamine (OAm)

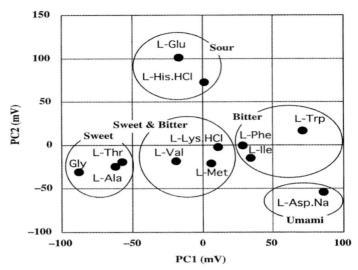

Figure 2.2-1. Discrimination of taste of amino acids using the taste sensor. Reprinted from Toko, K., Taste sensor, *Sens. Actuators B* **64** (2000) 205–215, with permission from Elsevier Science.

The sensitivity of the taste sensor was studied in aqueous solutions of five basic taste substances: salty (NaCl, KCl, KBr), sour (HCl, citric and acetic acids), bitter (quinine), umami (monosodium glutamate) and sweet (sucrose) [36, 37]. The taste sensor output exhibits different patterns for chemical substances which have different tastes, whereas it exhibits similar patterns for chemical substances with similar tastes. The sensor sensitivity (response slope) to sour and salty substances, ie, HCl, organic acids, NaCl, KCl, and KBr, was about 50–60 mV/pX. These values correspond to the typical values known for ISEs and may be explained by the sensitivity to pH, alkali metal cations and halogen anions. The sensitivity to glutamate was found for one sensor and the slope was about 13 mV/pX. Also, the sensitivity of some sensors to quinine hydrochloride with a response slope of about 50 mV/decade was demonstrated. The reported sensitivity to another alkaloid, caffeine, was low, about 5 mV/pX. The sensitivity to natural sweet substances (sucrose) was very low. In contrast, some sensors responded to the change in concentration of the artificial sweetener aspartame with a slope of about 40 mV/pX. Therefore, in the cases when the sugar concentration was crucial, an enzymatic glucose-selective sensor was used together with the taste sensor [38].

The sensitivity of the device to astringent and pungent substances was investigated [39, 40]. No sensitivity to pungent substances such as capsaicin, piperine, and allyl isothiocyanate was found. However, the taste sensor displayed a response to substances with an astringent taste: catechin, tannic acid, chlorogenic acid, and gallic acid. The astringency area was located between bitterness and sourness on principal component plots of the taste substances. Amino acids and dipeptides, which elicit complicated tastes from sour and bitter to sweet, were

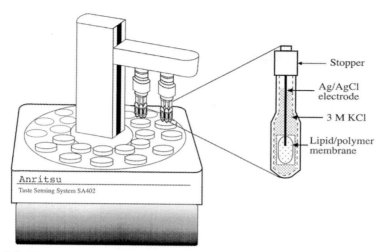

Figure 2.2-2. Taste-sensing system SA402, Anritsu. Reprinted from Toko, K., Taste sensor, *Sensors and Actuators B* **64** (2000) 205–215, with permission from Elsevier Science.

classified according to their tastes using the taste sensor [41–44]. A principal component analysis (PCA) score plot of amino acids is shown in Figure 2.2-2. The strength of bitterness of the amino acid L-tryptophan was estimated in terms of quinine concentration.

The sensitivity to taste substances of polymer membranes of the same compositions as described above was studied using impedance measurements [45, 46]. The impedance measurements were performed on thin-film membranes, which were prepared by dip-coating or by deposition of a Langmuir–Blodgett film. In contrast to the electric potential, the impedance of the membrane changed significantly in the presence of two umami substances simultaneously and sucrose. It was also found that impedance of the Langmuir–Blodgett polymer film increased in the presence of many bitter substances, which were both electrolytes and nonelectrolytes.

A miniaturized version of the taste sensor was developed on the basis of metal oxide semiconductor field effect transistor (MOSFET) technology [47]. A field effect transistor (FET) taste sensor was prepared by pasting membranes of the same compositions as above on the gate of a FET. Another method used for FET sensitive layer preparation involved the deposition of a dihexadecyl phosphate Langmuir–Blodgett film. A PVC polymer film was deposited on the MOSFET gate prior to Langmuir–Blodgett layer formation to enhance adhesion of the latter to the substrate. The FET taste sensor displayed the same sensitivity to taste substances as the conventional one, but the potential reproducibility was lower and the lifetime was shorter for the miniaturized device.

Sensors with the same polymer lipid membranes as the sensitive layer but employing other principles of signal transduction and detection have been developed and used in electronic tongue systems. A microsystem of light addressable potentiometric sensors (LAPS) with polymer lipid membranes was developed [48]. Owing to a new differential measuring procedure based on a time-sharing technique, the sensitivity of the sensor system was increased by two orders of magnitude in comparison with a conventional LAPS system. The response of the sensor system was studied in solutions of taste substances and it was found to be able to identify the sweet ones. A surface photovoltage (SPV) technique was used for signal detection from the taste sensor by the same group [49]. The sensitive layer was prepared by integrating polymer membranes containing active substances on to a semiconductor surface using the Langmuir–Blodgett technique. The surface potential change caused by the reactions with taste substances was detected by scanning a light beam across the semiconductor surface. The device displayed sensitivity to five basic taste substances and was able to recognize different commercial soft drinks.

An electronic tongue based on potentiometric sensors was developed in the Laboratory of Chemical Sensors of St. Petersburg University [50, 51]. The same group together with Italian colleagues suggested the term 'electronic tongue' referring to multisensor systems comprising cross-sensitive (partially selective) sensors and pattern recognition tools for data processing.

A large number of sensing materials of different natures from inorganic compositions to organic polymers have been used for sensor preparation. The membrane materials included chalcogenide glasses doped with different metals, PVC-based polymers containing various plasticizers and active substances such as ionophores, neutral carriers, metalloporphyrins, etc., and crystalline compositions [25–27]. The sensor arrays comprised from 10 to 30 sensors depending on the application. Different sensors with both a liquid inner filling and a solid inner contact have been manufactured and applied.

Two types of experimental set-up were used for measurements with the electronic tongue. Many experiments were performed in a 'static mode'. A schematic diagram and photograph of this system are shown in Figures 2.2-3 and 2.2-4, respectively. The potential values for each sensor were measured versus an Ag/AgCl reference electrode using a multichannel measuring device with a high input impedance controlled by a PC. The experimental data were written to computer files. A multisensor flow injection system was also developed. The measuring cells of different sizes and configurations were manufactured and the system could comprise from 3 to 11 sensors, which were made in special small bodies for this purpose. The flow injection multisensor system allows automated sampling and calibration using a multichannel actuator. The sample volume in this case could be as small as 50 µL, usually ranging from 150 to 500 µL. The overall performances of static and flow sensor arrays systems were comparable.

Microsensors based on thin films of chalcogenide glass were prepared using a new technology, pulse laser deposition (PLD) [52]. The resulting thin films had the same composition as corresponding bulk glasses. Thin-film microsensors showed similar electrochemical properties, such as sensitivity and selectivity, to conventional sensors with bulk membranes. The lifetime of microsensors was found to be about 2 years. Thus, PLD seems to be a promising tool for sensor miniaturization and the preparation of microsensor arrays for the electronic tongue.

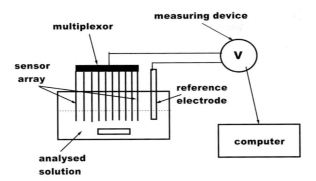

Figure 2.2-3. Schematic diagram of potentiometric electronic tongue.

Figure 2.2-4. Laboratory prototype of potentiometric electronic tongue.

Originally, the electronic tongue was intended for the analysis of very different types of sample including ground and natural waters, industrial solutions, medicinal solutions, foodstuffs, etc. Hence the sensitivity of sensors to numerous inorganic and organic substances was studied prior to their application, special attention being paid to sensor cross-sensitivity and response reproducibility. The sensitivity of the sensors with a chalcogenide glass membrane to heavy metal cations (eg, copper, lead, cadmium, zinc, iron, and uranium in different oxidation states, etc.) was investigated [53–57]. PVC-based sensors displayed sensitivity to both alkali and alkaline earth and heavy metal cations and inorganic anions such as chloride, sulfate, nitrate, nitrite, carbonate, etc. [27, 58]. The sensitivity of about 40 different sensors to organic taste substances present in many foodstuffs and responsible for different aspects of food quality and flavor was investigated

[59]. The taste substances were organic acids, alcohols, ethers, aldehydes, phenols, amines, alkaloids, etc. Also, the sensitivity to typical organic 'basic taste' substances, such as quinine, urea, and glutamate, was evaluated.

There is experimental evidence that sensor arrays display some new specific features in comparison with the best available discrete selective sensors. Application of a cross-sensitive sensor array permitted a lower detection limit and increased selectivity [60]. Another interesting feature of the sensor arrays is the possibility of making potentiometric measurements without a reference electrode. Potential differences in this case are measured between all sensor pairs in the array [61]. The resulting potential differences were considered during the data processing and redundant values were rejected. Two PCA score plots displaying the electronic tongue data measured with and without a reference electrode are shown in Figure 2.2-5 a and b. Although the absolute positions of the points are different on these two plots, the instrument can still easily distinguish the difference between classes. Therefore, the discrimination abilities of the electronic tongue do not depend on the data used (measurements with or without a reference electrode), at least for selected applications.

A different analytical technique, voltammetry, was used for electronic tongue development by Swedish scientists at Linkoping University [62–66]. The same general idea of combining non-specific and partly overlapping sensor signals with pattern recognition tools was realized in the voltammetric electronic tongue. In this case, nonspecific signal patterns are produced by an array of working electrodes made from different noble metals. Up to six metal electrodes, made of platinum, gold, iridium, palladium, rhodium, and rhenium, were combined and used for different tasks. A sketch of the voltammetric sensor array is shown in Figure 2.2-6. The measurements with the array were performed in a three-electrode scheme including an Ag/AgCl reference electrode, a stainless-steel auxiliary electrode and an array of noble metal working electrodes. Two types of voltammetric techniques were investigated: small amplitude pulse voltammetry (SAPV) and large amplitude pulse voltammetry (LAPV). LAPV was chosen for further analytical application of the electronic tongue. Different voltage steps ranging from 15 to 100 mV were used. Thus, a number of data points for each electrode were collected. The most informative data points were chosen during the data processing and used for calibration and recognition. A hybrid electronic tongue combining voltammetry with six working electrodes, three ion-selective electrodes (for pH, chloride, and carbon dioxide), conductivity and temperature sensors was also suggested [64]. A schematic diagram of the hybrid electronic tongue is shown in Figure 2.2-7.

An electronic tongue aiming at mimicking biological sensory systems was developed at the University of Texas [67]. Sensors mimicking taste buds were prepared from poly(ethylene glycol)–polystyrene (PEG–PS) resin beads that were derivatized with a variety of indicator molecules. The sensors were fluorescein sensitive to pH, *o*-cresolphthalein complexone sensitive to Ca^{2+} and pH, alizarin complexone sensitive to Ce^{3+}, Ca^{2+}, and pH, and finally a boronic ester of resorufin-derivatized galactose sensitive to simple sugars. The sensor responses were measured versus a reference sensor, which was simply a resin bead with the ter-

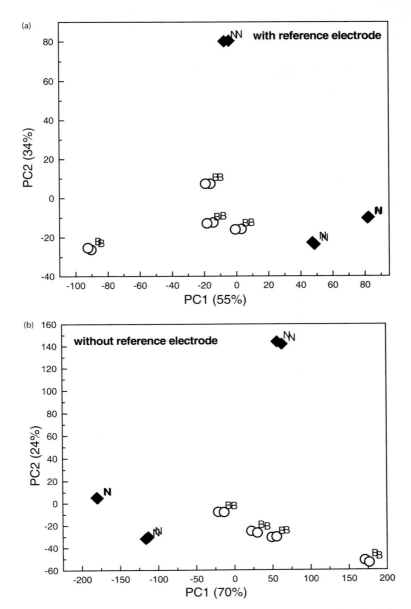

Figure 2.2-5. Discrimination of urine samples containing blood and from healthy patients using the electronic tongue: ○ urine samples containing blood; ◆ urine samples from healthy subjects. (a) Data with references electrode; (b) data without reference electrode.

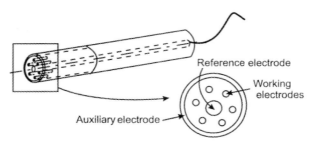

Figure 2.2-6. Electrode arrangement in the voltammetric electronic tongue. Reprinted from Winquist, F., Holmin, S., Krantz-Rülcker, C., Wide, P., Lundström, I., A hybrid electronic tongue, *Analytica Chimica Acta* **406** (2000) 147–157, with permission from Elsevier Science.

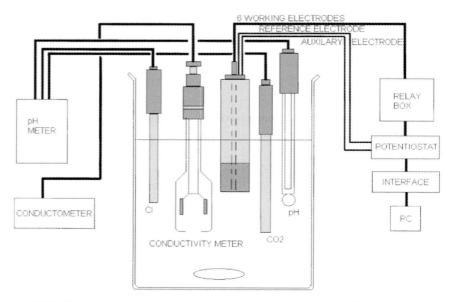

Figure 2.2-7. The hybrid set-up of the electronic tongue including the voltammetric, potentiometric, and conductometric electrodes. Reprinted from Winquist, F., Holmin, S., Krantz-Rülcker, C., Wide, P., Lundström, I., A hybrid electronic tongue, *Analytica Chimica Acta,* **406** (2000) 147–157, with permission from Elsevier Science.

minal amines acetylated. To mimic the cavities in which natural taste buds reside, the resin beads were positioned within micromachined wells formed in Si/Si₃N₄ wafers, thus confining the beads to individually addressable positions on a multicomponent chip. The size of the wells was chosen so that they held the beads in swollen and unswollen states. Signal transduction was accomplished by the analysis of the light absorption changes of the beads using a charge-coupled device that was interfaced with the sensor array. The sensor sensitivity was tested in solutions containing Ca^{2+}, Ce^{3+}, and their mixture at different pH levels. The aim of this work was a proof of concept of a sensor system design mimicking a biological tongue, which explains the somewhat strange choice and limited number of analytes and the absence of information about application of the device to the analysis of any real-world samples.

A sensing microsystem based on shear horizontal surface acoustic wave (SH-SAW) devices on $36°$ rotated Y-cut X-propagating $LiTaO_3$ was applied for the identification of liquids [68]. The system consisted of three SH-SAW devices with three different center frequencies of 30, 50, and 100 MHz.

More references to and mentions of electronic tongue devices may be found in the literature. However, the present description is restricted to the most reliable and well-documented results regarding electronic tongue sensor systems and their applications.

2.2.4   Applications of Electronic Tongues

The main application area of electronic tongue systems is the analysis of foodstuffs and beverages in particular. The analysis of foodstuffs included different tasks: discrimination between different sorts, brands, and products of different quality, their classification, and quantitative determination of the content of various compounds. The most exciting application area of electronic tongues is taste quantification, which is understood as the assessment of taste (flavor) characteristics of a product using an electronic tongue and correlation of its response to human sensory perception. Since the different types of analysis were in many cases performed simultaneously and during the same experimental sessions, the discussion of electronic tongue applications will be arranged according the media analyzed.

2.2.4.1   Model Experiments with Taste Substances

Let us begin with the discussion of model experiments, which were aimed at establishing taste quantification using an electronic tongue. In these studies the measurements were made in individual and mixed solutions of taste substances.

Taste quantification using the potentiometric electronic tongue was performed in solutions containing three taste substances: NaCl (salty), lactic acid (sour), and L-leucine (bitter), and their binary mixtures [59]. A PCA score plot of recognition of taste substances binary mixtures is shown in Figure 2.2-8. After the electronic tongue calibration in individual solutions, the presence and intensity of each taste in binary mixtures were correctly predicted (Table 2.2-3).

An attempt to build a taste map using the taste sensor and thereby to express the tastes of foodstuffs by combinations of basic tastes was performed using the taste

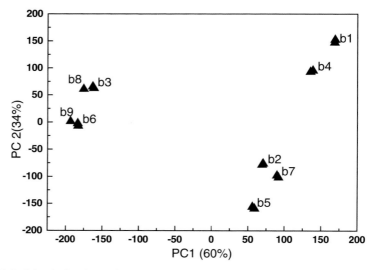

Figure 2.2-8. Discrimination of taste substance binary mixtures by the electronic tongue. b1, b4, mixtures of lactic acid and L-leucine (sour and bitter taste); b2, b5, b7, mixtures of sodium chloride and lactic acid (sour and salty taste); b3, b6, b8, b9, mixtures of sodium chloride and L-leucine (bitter and salty taste).

Table 2.2-3. Prediction of taste intensity in binary mixtures of taste substances using the electronic tongue

Sample	Sourness		Bitterness		Saltiness	
	Real	Predicted	Real	Predicted	Real	Predicted
B1	low	low	low	low	0	0
B2	low	medium	0	0	low	medium
B3	0	0	low	0	low	medium
B4	high	high	high	high	0	0
B5	high	high	0	0	high	high
B6	0	0	high	low	high	high
B7	high	high	0	0	low	medium
B8	0	0	high	medium	low	low
B9	0	0	low	low	high	high

sensor [69]. For this purpose 256 solutions containing four basic taste substances, NaCl, HCl, quinine, and sucrose, at four concentration levels were prepared and measured. The dependence of each sensor potential on four component concentrations was approximated by a quadratic function. Then measurements of commercial soft drinks were performed and the resulting pattern was used to calculate the quan-

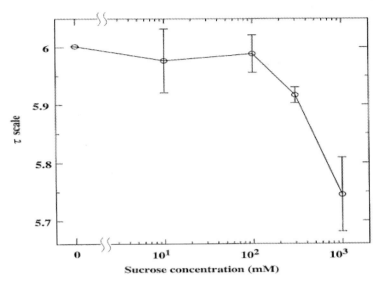

Figure 2.2-9. Suppression of bitterness of quinine by sucrose expressed by the τ scale. Reprinted from Toko, K., Taste sensor, *Sens. Actuators B* **64** (2000) 205–215, with permission from Elsevier Science.

tities of basic taste substances necessary to produce the same taste. A drink made with calculated contents of basic substances appeared to have the same taste as the commercial product according to human perception. However, the possibility of modeling the taste by a combination of four basic tastes was successfully demonstrated on only one sample of a commercial soft drink.

The suppression of certain tastes perceived by humans in the presence of the other substances is well known. The suppression of bitter taste by sweet substances, which is often used to mask the bitter taste of drugs, was studied using a taste sensor [70, 71]. The response of the taste sensor to quinine was expressed using the τ scale that is used to describe taste strength. The human response to taste stimuli and the response of a potentiometric chemical sensor depend linearly on the logarithm of substance concentration. Thus, a linear relationship between the strength of bitterness τ and first principal component calculated from the taste sensor data was established. The measurements in individual quinine solutions were used in this case. The degree of bitterness calculated from the taste sensor response in solutions containing a constant quinine concentration of 1 mM and changing sucrose concentration is shown in Figure 2.2-9. The predicted degree of bitterness dropped significantly at a sucrose concentration of 1 M. The same experiment was repeated with artificial sweet substances, ie, phospholipids, which are used in pharmacology for masking the bitter taste of drugs. It was found that four basic taste substances, sour, salty, bitter, and sweet, mutually suppressed the increase in the taste sensor response [72]. The suppression of saltiness by monosodium glutamate, which elicits an umami taste, was detected using the taste sensor, quantification being done by PCA [73].

2.2.4.2   Analysis of Beverages

Pretreatment of the sensors of electronic tongues before measurements was often performed with the aim of reducing drift and enhancing reproducibility. The procedures varied from system to system. A potentiometric electronic tongue was washed in distilled water until the sensors reached a steady potential. The working electrodes of the voltammetric electronic tongue were first polished and then cleaned by applying subsequently positive and negative potentials of 2 V. Measurements with the taste sensor in beer, milk, and saké were made after conditioning the taste sensor in a corresponding standard beverage. The potential pattern of the taste sensor in the standard beverage was considered as the zero level during data processing. The beverages used as the standard ones were chosen at random in this case.

2.2.4.2.1 Mineral Water

A large number of beverages and liquid foods have been analyzed using electronic tongue systems. Mineral waters represent relatively simple and, hence, ideal samples for analysis using the sensor systems. Discrimination and classification of natural and fake Georgian and Italian produced mineral waters were performed using a potentiometric electronic tongue [74–77]. Discrimination was made using PCA and a Kohonen map and classification was performed using SIMCA and a back-propagation neural net. A comparison of the performances of two parallel sensor arrays was also made. Calibration was done using the data produced by one of the arrays and the prediction of test sample membership was made using the results of the measurements made with the other. All the samples were classified correctly. Representation of the response of the two arrays in mineral water made by a Kohonen neural map is shown in Figure 2.2-10. Thus, the reproducibility of the sensor properties is good enough to use the arrays interchangeably. Italian mineral water samples and tap water were discriminated using the electronic tongue. Simultaneously, quantitative determination of the content of ions such as sodium, potassium, chloride, fluoride, sulfate, bicarbonate, and nitrate was performed. The electronic tongue output also correlates with parameters such as conductivity and dry matter residual, which are related to total salt content in mineral water. The results of the quantitative analysis of mineral water are shown in Table 2.2-4. The ability of the electronic tongue to detect foreign substances in mineral water was evaluated. For this purpose, two samples of mineral water were deliberately contaminated with organic matter (a piece of fruit). A PCA score plot of measurements made on the following day shows that contaminated samples could be easily distinguished from pure samples (Figure 2.2-11).

Classification of mineral water samples according to their hardness was performed using a taste sensor [78]. The measurements were performed on 41 kinds of commercial mineral waters. After data processing by PCA, it appeared that

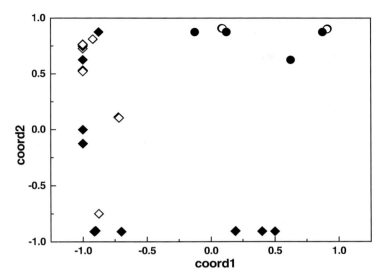

Figure 2.2-10. Representation of the response of two sensor arrays in mineral waters made by a Kohonen neural map. Diamonds, natural mineral waters; circles, counterfeited mineral waters; solid symbols, response of the first sensor array used for calibrations; open symbols, response of the second sensor arrays used as test.

the first principal component correlates with water hardness. Sensory evaluation of seven mineral water samples was performed. However, mineral water taste does not vary significantly in human perception and the taste assessment by the sensory panel was neither meaningful nor reproducible. As a result, a correlation of the taste sensor response with human perception was not found in this case.

2.2.4.2.2 Coffee

Recognition of different sorts of coffee using the electronic tongue and especially the correlation of instrument output with human perception of coffee flavor are challenging and important tasks. All commercial brands of coffee are blends of different sorts of coffee beans. The quality of harvests differs from year to year but the same coffee brands must be prepared each year with a reproducible and familiar characteristic flavor. This procedure is performed by taste panels and, hence, is time consuming and expensive. The possibility of partly replacing humans in the coffee industry could have a large practical impact. Therefore, numerous attempts to use sensor systems for coffee analysis were carried out.

Measurements with a taste sensor were performed on 10 brands of coffee of different origins, one of which was used as the standard [79]. The measurements were carried out with coffee brews at 60 °C. The response of the sensor containing oleic acid was correlated with coffee acidity estimated by tasters with a correlation coefficient of 0.98. The correlation between the response of the sensor

Table 2.2-4. Results of quantitative analysis of mineral waters using the electronic tongue

Water	Cond.[a] real (µS/cm)	Cond. found (µS/cm)	Standard deviation (µS/cm)	Mean error (%)
Levissima	107.5	110	9	5
Fuiggi	156	155	1	6
Uliveto	1388	1414	69	4
Sangemini	1333	1302	204	14
Ferrarelle	1800	1792	120	5
S. Pellegrino	1306	1516	112	16
Tap	−[b]	976	116	−
	Resid.[c] real (g/L)	Resid. found (g/L)	Standard deviation (g/L)	Mean error (%)
Levissima	0.0735	0.075	0.006	8
Fuiggi	0.1065	0.105	0.002	2
Uliveto	0.986	1.03	0.05	5
Sangemini	0.9550	0.9	0.2	15
Ferrarelle	1.283	1.28	0.08	5
S. Pellegrino	1.109	1.10	0.09	6
Tap	−[b]	0.70	0.08	−
	SiO_2 real content (g/L)	SiO_2 found (g/L)	Standard deviation (g/L)	Mean error (%)
Levissima	5.8×10^{-3}	5.9×10^{-3}	4×10^{-4}	5
Fuiggi	1.88×10^{-2}	1.97×10^{-2}	6×10^{-4}	5
Uliveto	7.0×10^{-3}	7.1×10^{-3}	2×10^{-4}	3
Sangemini	2.45×10^{-2}	2.2×10^{-2}	2×10^{-3}	10
Ferrarelle	8.3×10^{-2}	8.5×10^{-2}	7×10^{-3}	7
S. Pellegrino	9×10^{-3}	7.9×10^{-3}	5×10^{-4}	12
Tap	−[b]	5.5×10^{-2}	7×10^{-3}	−

[a] Cond., electrical conductivity.
[b] For tap water, quantitative analysis results are not available.
[c] Resid., dry residue (after evaporation of all water).

containing dioctyl phosphate and trioctylmethylammonium chloride and coffee bitterness estimated by tasters was also good ($R=0.94$).

Coffee recognition was performed using a potentiometric electronic tongue [77]. Ten coffee samples were analyzed, consisting of seven individual sorts and three commercial brands. The measurements were performed on coffee brews which were prepared using a weighed amount of coffee and an exact water volume and cooled to room temperature. The electronic tongue was able to distinguish all coffee samples. The experiments made with different coffee concentrations and water compositions (distilled, distilled with sodium chloride and tap water) showed that the coffee samples were correctly distinguished in all cases.

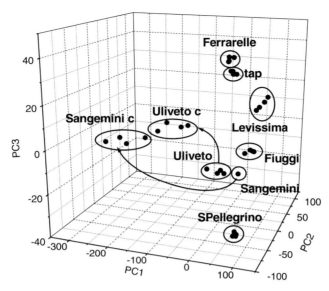

Figure 2.2-11. PCA score plot for seven Italian mineral waters including two contaminated with organic matter (Uliveto C and Sangemini C).

Coffee taste assessments made by a professional sensory panel were obtained from the manufacturer along with the coffee samples. Four parameters were evaluated: flavor, acidity, body, and smell, which obviously characterize the taste, odor and flavor of the coffee. A PCA score plot of coffee sample recognition together with direction of change of coffee flavor parameters is shown in Figure 2.2-12. It was found that the first PC correlates with the flavor and smell of coffee. Then multivariate calibration was performed using a back-propagation neural network, taste panel scores being used as the reference data. Subsequently, the electronic tongue could correctly predict the sensory assessment values of all four flavor parameters.

2.2.4.2.3 *Fruit Juices*

Fruit juices are popular analytes for electronic tongue sensor systems. The potentiometric electronic tongue was applied to discriminate orange juices and monitor juice spoilage processes [80]. The ability of the electronic tongue to trace the spoilage of juice was studied on grape and orange juices. In the first experiment, two identical packages of grape juice were opened simultaneously and measured every hour during the first 4 h and then twice per day for a week. One bottle with juice was stored in a refrigerator and the other at room temperature. The electronic tongue could monitor the spoilage process of the juice, which was different for samples stored at different temperatures. For an advanced follow-up of

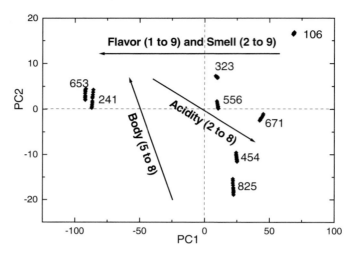

Figure 2.2-12. PCA score plot of different sorts of coffee.

juice spoilage, the second experiment was performed. The measurements were made in orange juice every hour for 5 h after opening and subsequently six times per day for a week. The juice was kept in a refrigerator. After experimental data processing by PCA, the juice spoilage process can be traced. Fast changes, which were probably related to juice oxidation by air, were observed during the first few hours after package opening and were followed by slow evolution during the 5 days. When finally the juice was spoiled the corresponding points on the PCA score plot were distinctly separated from the others. Another experiment with the potentiometric electronic tongue was performed on 11 samples of juices produced by the same company, including seven samples of orange juice and four samples of orange juice with lemon additive. Of the seven orange juice samples, the storage time had expired for three of them [81]. The electronic tongue was capable of distinguishing orange juices with lemon additive from pure orange juices and also edible juice samples from those for which the storage time had expired.

The potentiometric electronic tongue was also applied to the analysis of tomato juice [82]. Eleven kinds of tomatoes, which were grown in experimental orchards of an agricultural institute and harvested at two different times, were studied. The measurements with the electronic tongue were carried out in clarified juice with solids removed by centrifugation. Chemical analysis by conventional analytical techniques (HPLC, AAS, GS, etc.) and sensory evaluation by a trained panel were performed and these data were used for calibration of the electronic tongue. It was found that the sensor system could recognize tomato juice samples made from different varieties and from the same variety at different stages of ripening. The concentrations of inorganic substances (K, Na, Mg, Ca, phosphate), organic acids (malic and citric) and UMP were determined with an average precision 10–15%. The electronic tongue output was correlated with human

sensory perception and used to predict the scores of the panellists in the evaluation of acidity, sweetness, bitterness, saltiness, umami, and sharpness of tomatoes. A good correlation between the electronic tongue and sensory panel was
observed. All three types of analysis were performed simultaneously on the basis
of the same experimental data set.

A voltammetric electronic tongue with two working electrodes (platinum and
gold) was applied to discriminate different beverages, including nine brands of
orange juices, both plain and concentrated, two types of orange soft drinks, apple
juice, and pasteurized milk. The concentrated juices were diluted before measurements as recommended on the package. It was found that the device was able to
distinguish different types of beverages after data processing by PCA [62]. The
same set-up was applied to study the process of beverage aging. The measurements were made in milk and orange juice kept at room temperature for 15 and
20 h, respectively. The electronic tongue could track the changes in beverages related, most probably, to their oxidation, evaporation of volatile compounds, etc.
(Figure 2.2-13 a and b).

A system consisting of three SH-SAW devices with three different center frequencies was used for measurements in fruit juices. Classification of eleven
kinds of fruit juices was achieved using this system, the experimental data being
processed by PCA and discriminant analysis [68].

2.2.4.2.4 Soft Drinks

The potentiometric electronic tongue was used for discriminating regular and diet
sodas and experimental recipes [77]. The regular and diet sodas and the experimental samples differed mainly in the type of sweetener used. Eight samples
were studied: regular and diet Pepsi Cola and Coca Cola and four experimental
recipes. The electronic tongue was able to recognize all samples after data processing by PCA. The regular and diet sodas formed two groups and the experimental mixtures lay between them. The degree of 'dietness' of sodas assessed by
a trained sensory panel was obtained from the manufacturer along with samples.
After calibration (PLS), the electronic tongue could correctly predict the degree
of 'dietness' for commercial drinks and experimental mixtures.

2.2.4.2.5 Milk and Dairy Products

Another widespread application area of electronic tongues is the analysis of fresh
and fermented milk. The most important practical tasks are the discrimination of
milk that has undergone different heat treatments, since the latter affects both
milk flavor and nutritive value, and also monitoring of the milk souring process
and bacterial growth.

The taste sensor was applied for measurements in milk focusing on the recognition of samples that had undergone different heat treatments and correlation of
the system output with human perception [83]. The measurements were per-

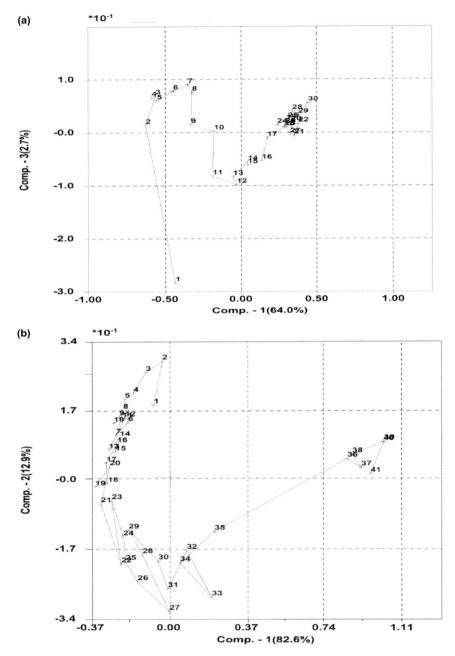

Figure 2.2-13. Aging process for (a) orange juice and (b) milk as represented by a score plot. The time between each sample is 30 min. Reprinted from Winquist, F., Wide, P., Lundström, I., An electronic tongue based on voltammetry, *Analytica Chimica Acta* **357** (1997) 21–31, with permission from Elsevier Science.

formed on seven milk samples. The taste sensor was conditioned before measurements for approximately 1 month in a milk sample chosen as the standard sample. Four milk characteristics were evaluated: richness, deliciousness, and cooked flavor estimated by tasters and whey-protein nitrogen index (WPNI), which defines the degree of protein denaturation due to the heating and which is determined by infrared spectroscopy. It was found that the response of the sensor containing trioctylmethylammonium chloride was correlated with the richness of milk taste with a correlation coefficient of 0.885. The response of the sensor containing decyl alcohol was correlated with WPNI with a correlation coefficient of 0.953. An attempt to distinguish milk samples which had been homogenized under different pressures and thus contained fat globules of different size was performed [85]. However, the sensor responses were only slightly different in the milk samples and the potential changes were <0.5 mV.

Milk samples that had undergone different heat treatments (ultra-high temperature (UHT) and pasteurized) from different manufacturers stored at room temperature and in a refrigerator were measured using a potentiometric electronic tongue [86, 87]. The device was able to discriminate UHT and pasteurized milk samples and also the same type of milk produced by different manufacturers. The milk souring process, which was different for milk stored at different temperatures, could be easily detected and traced by the system. The discriminating ability of the electronic tongue is illustrated in Figure 2.2-14. An integrated sensor system comprising an electronic tongue and an electronic nose was used for measurements in this experiment. Fusion of gas and liquid sensor systems will be discussed in greater detail in Section 2.2.4.4.

A voltammetric electronic tongue with five working electrodes (platinum, gold, iridium, palladium, and rhodium) was applied for monitoring the milk sour-

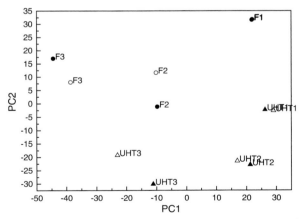

Figure 2.2-14. Discrimination of milk samples that had undergone different temperature treatment and monitoring of the milk souring process with the potentiometric electronic tongue. Circles, fresh (pasteurized) milk samples; triangles, UHT-treated milk samples; open symbols, milk samples stored in a refrigerator; solid symbols, milk samples stored at room temperature; index, day of measurements.

ing process and bacterial growth estimation [63]. The measurements were performed on 11 samples of milk immediately after package opening and then at intervals of about 30 min for a maximum of 18 h. A bacteriostatic agent, sodium
azide, was added to two milk samples to prevent bacterial growth. Bacterial colo-
ny counts in milk were checked every 2 h using the dip-slide test and the results
were subsequently used as reference values for calibration. Multivariate calibration was performed using PLS regression and a back-propagation ANN. The results of prediction were acceptable in most cases. However, it is not clear what
the mechanism of the sensitivity of the voltammetric electronic tongue to bacteria is.

Six different types of fermented milk were analyzed using a voltammetric electronic tongue complemented by three potentiometric, conductivity, and temperature
sensors. The fermented milk samples differed in the microorganisms used for their
preparation. It was demonstrated that the combination of signals of different nature,
eg, voltammetric, potentiometric, and conductometric improves the discrimination
abilities of the device. A hybrid electronic tongue could distinguish almost all fermented milk samples (Figure 2.2-15). The data were processed using PCA and a
back-propagation ANN, the latter being used for classification. Almost all milk
samples were assigned to a correct class (Figure 2.2-16).

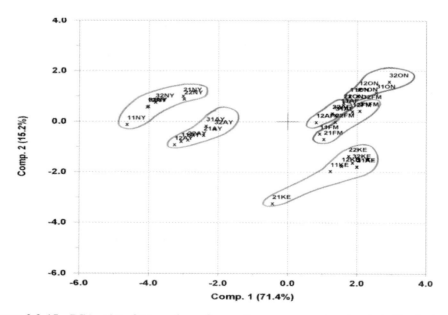

Figure 2.2-15. PCA plot from values from all measurements: the hybrid electronic
tongue. Reprinted from Winquist, F., Holmin, S., Krantz-Rülcker, C., Wide, P., Lundström, I., A hybrid electronic tongue, *Analytica Chimica Acta* **406** (2000) 147–157, with
permission from Elsevier Science.

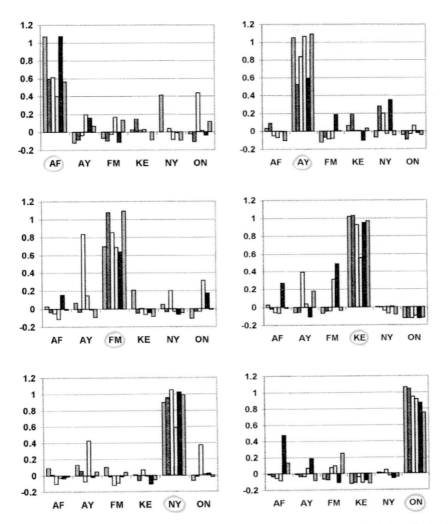

Figure 2.2-16. Predicted versus true values for the training with the ANN for all six different samples. The true class in each prediction is encircled. Reprinted from Winquist, F., Holmin, S., Krantz-Rülcker, C., Wide, P., Lundström, I., A hybrid electronic tongue, *Analytica Chimica Acta* **406** (2000) 147–157, with permission from Elsevier Science.

2.2.4.2.6 Alcoholic Beverages

The potentiometric electronic tongue was applied to Italian red wine analysis [74, 75, 87]. The measurements were performed on 20 samples of wine of the same type (Barbera) and the same vintage but from different vineyards. Measurements were also carried out on two samples of Castelli Romani wine, red and white. It was found that the electronic tongue could distinguish wines of differ-

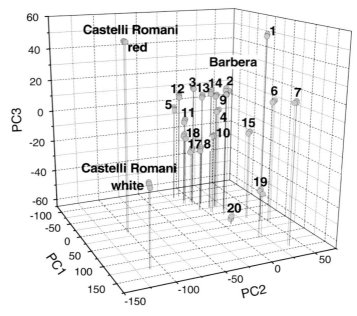

Figure 2.2-17. Discrimination of Castelli Romani and Barbera Italian wines by the potentiometric electronic tongue.

ent types and also all 20 samples of Barbera wine (Figure 2.2-17). Results of the
standard chemical analysis and flavor quality assessments of the wine samples
made by a taste panel were provided by the Italian Enological Institute. This
data was used as reference values for calibration of the electronic tongue. Multivariate calibration for quantitative determination of wine components was performed using PLS. The electronic tongue could predict the content of alcohol
and organic acids in wine and also the total and volatile acidity and pH with
average errors of about 10%. The taste panel evaluated wine using 14 parameters
describing different aspects of its taste, odor and color. Multivariate calibration
for determination of wine flavor parameters was performed with a back-propagation ANN. After the calibration the electronic tongue could predict all wine flavor scores with average errors of about 15%.

Thirty-six brands of beer, both Japanese produced and imported, were measured using a taste sensor [69, 88]. After data processing by PCA, it appeared
that the first principal component correlates with beer taste richness/lightness and
the second principal component correlates with mild/sharp touch taste of beer.
The correlation between taste sensor output and sensory estimates such as 'bitter', 'rich taste', and 'sharp touch taste' was observed. The sensitivity to ethanol
concentration and pH was also detected (Figure 2.2-18).

The above taste sensor developed in Japan was applied to the analysis of two
traditional Japanese foodstuffs, saké and miso (soybean paste). An attempt to
measure the concentration of ethanol and total acidity changes during saké fer-

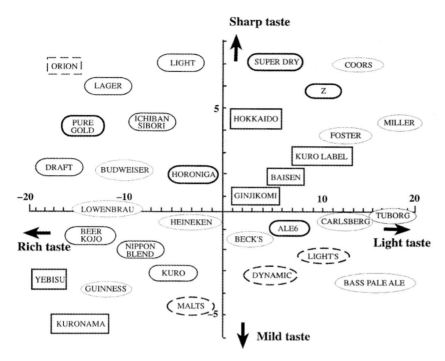

Figure 2.2-18. Taste map of beer. Symbols enclosing the brand of beer, such as dotted el-lipses and squares, represent beer produced by different companies in Japan, and thin sol-id ellipses represent beer from other countries. Reprinted from Toko, K., Taste sensor, *Sens. Actuators B* **64** (2000) 205–215, with permission from Elsevier Science.

mentation using the taste sensor was performed [89]. Four kinds of saké from the same brewery were measured, an ordinary saké being used as reference. The calibration model for titratable acidity determination was made by multiple linear regression using the outputs of two sensors. The correlation between titratable acidity and sensor responses was as high as 0.99. The results of ethanol content determination in saké by the taste sensor were in good agreement with the values obtained using gas chromatography [90]. The taste of saké was evaluated using the taste sensor together with an enzymatic glucose-selective sensor [38]. In con-trast to other applications, the artificial mixtures were used as standard solutions instead of a selected type of beverage. Three standard solutions of complex com-positions were prepared that contained different concentrations of glucose, etha-nol, succinic acid, lactic acid, sodium succinate, tyrosol, monosodium glutamate, glycine, alanine, sodium chloride, and potassium and calcium hydrophosphates. Quantification of saké taste was performed using two parameters: acidity and su-gar content. Two sets of standard solutions were measured, the first one with a variable content of succinic acid and the second with a changing glucose con-tent. Then the taste sensor output and glucose sensor response were transformed to produce two new variables, corresponding to succinic acid and glucose content, respectively. The measurements in 17 brands of saké were projected on this taste map. However, no comparison of saké taste with human sensory perception was performed in this study.

2.2.4.3   Analysis of Vegetable Oils

An untypical foodstuff for chemical sensor analysis, vegetable oil, was analyzed using the potentiometric electronic tongue [91]. Since the oil is not conductive, a special procedure of oil extraction by an organic solvent was elaborated, the resulting extract being used for the measurements. Soybean, rapeseed, and corn refined and olive mixed with vegetable, olive extra virgin, and sunflower unrefined oils were studied. The electronic tongue could distinguish all vegetable oils, including oils very close in composition such as olive extra virgin and olive refined. The electronic tongue also reliably detected the rancidity in oils.

2.2.4.4   Analysis of Non-Liquid Food

2.2.4.4.1 Fruits and Vegetables

Although electronic tongues were primarily designed for the analysis of liquids, they could also be applied for measurements in suspensions, purées, and other water–solid mixtures or homogenates. A sample preparation is a necessary step in this case and it can be performed in different ways. Products with a high water content such as fruits and some vegetables can be crushed and the measurements can be performed on the resulting pulp. Other products with lower water content, eg, flesh food, should first be minced and then mixed with distilled water.

Recognition of different tomato varieties was performed with a taste sensor [92]. The measurements were performed on crushed tomato pulp. For quantification of tomato taste, the taste sensor was first calibrated with canned tomato juice, to which four basic taste substances, NaCl, citric acid, monosodium glutamate, and glucose, were added. The experimental data were processed by PCA and then the measurements on several tomato varieties (without additives) were projected on to the principal axes, which represented a kind of 'taste map'. The resulting taste assessment agreed well with human perception.

A taste sensor was applied to measurements on another non-liquid foodstuff, miso, which is Japanese fermented soybean paste [93]. Chemical parameters (eg, titratable acidity) during the miso fermentation process are measured by conventional analytical methods, while ripeness and taste quality are estimated by humans. In the present study the possibility of replacing part of the routine analysis by simpler measurements using a sensor system was evaluated. It was found that

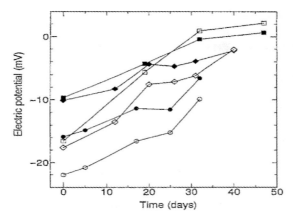

Figure 2.2-19. Changes in response of sensors 1 and 8 for komi-miso, mugi-miso, and awase-miso with length of fermentation. Circles denote changes for komi-miso, squares for mugi-miso and diamonds for awase-miso; solid symbols denote changes of sensor 1 and open symbols for sensor 8. Reprinted from Imamura, T., Toko, K., Yanagizawa, S., Kume, T., Monitoring of fermentation process of miso (soybean paste) using multichannel taste sensor, *Sens. Actuators B* **37** (1996) 179–185, with permission from Elsevier Science.

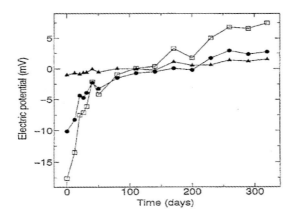

Figure 2.2-20. Changes in response of sensors 1, 8 and 15 for awase-miso with number of days of fermentation and storage. Reprinted from Imamura, T., Toko, K., Yanagizawa, S., Kume, T., Monitoring of fermentation process of miso (soybean paste) using multichannel taste sensor, *Sens. Actuators B* **37** (1996) 179–185, with permission from Elsevier Science.

the taste sensor output changed linearly with time of miso fermentation (Figure 2.2-19) whereas changes in sensor responses in ripe miso during the storage were smaller (Figure 2.2-20). The responses of the sensors with membranes containing dioctyl phosphate and oleylamine were correlated with total acidity of miso, the correlation coefficients being 0.87 and 0.88, respectively.

2.2.4.4.2 Flesh Food

The applicability of the electronic tongue to flesh food analysis was demonstrated on the examples of fish [77] and pork liver [94] recognition. The measurements were made in a homogenate prepared by stirring chopped fish with distilled water. It was found that the system was capable of distinguishing between a sample of freshwater fish and two samples of seawater fish. The measurements were performed on fish samples which had been stored in a freezer and at room temperature. The electronic tongue could easily detect and monitor fish spoilage (Figure 2.2-21).

Twelve samples of pork liver taken from animals of the same gender and age and similar breeding conditions were evaluated using the electronic tongue. Sample preparation was performed in the same way as for fish. Liver samples from different animals were reliably discriminated by the instrument. Furthermore, the liver from healthy animals was clearly separated from the liver from sick animals. Also, it was found that the electronic tongue could distinguish between samples of pork liver from animals fed with an anti-stress drug before slaughter and control animals which did not receive any medication.

2.2.4.5   Fusion of Sensor Systems for Gas and Liquid Analysis

In humans, the senses of olfaction and taste always cooperate and influence each other, especially in food flavor evaluation. Although the sensitivity of electronic and biological noses and tongues differ significantly in terms of the number of substances which can be identified and in the sensitivity thresholds, the idea of

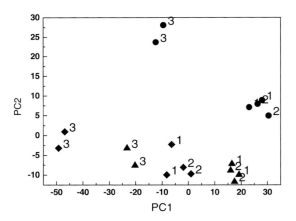

Figure 2.2-21. Discrimination of fresh and spoiled samples of seawater and freshwater fish by the potentiometric electronic tongue. ● Freshwater perch; ◆ haddock; ▲ cod; 1, 2, and 3 denote different days of measurements.

using an integrated sensor system, which would comprise both an electronic nose and an electronic tongue, sounds natural. Consequently, the possibility of using an integrated system for the analysis of foodstuffs, in particular, has been explored by many researchers.

The application of an integrated sensor system gives rise to the problem of combining data of different origins. Different approaches have been used for the fusion of electronic nose and tongue data. The simplest one assumes merging of two matrices of normalized data. The resulting matrix should have a number of rows equal to the number of samples and a number of columns equal to the number of sensors in the electronic nose and tongue. The data fusion may also be performed at higher levels. This means that the data from the electronic nose and tongue are analyzed separately and the results are combined afterwards. One method [85] is to apply, eg, PCA to nose and tongue data separately and then to consider together the most significant principal components. A more sophisticated approach has been suggested [65]. In this case, a hierarchical system of neural network-based classifiers was formed, some of them being trained with the electronic nose data and others with the electronic tongue data. A gating network that combines outputs of all classifiers gave the final results. Different applications of the integrated electronic nose–electronic tongue systems are discussed below.

A combination of an electronic nose and a voltammetric electronic tongue was applied to beverage classification [65, 66]. The electronic nose included 10 MOSFET sensors with the gates of thin catalytically active metals such as platinum, iridium, and palladium and four metal oxide (Taguchi) sensors. The electronic tongue consisted of six metal working electrodes (see above). The measurements were carried out with both devices on samples of three juices: apple, orange, and pineapple. The data were processed using PCA and PLS, the latter being used to build a classification model. It was found by PCA that using the electronic nose data alone the apple and pineapple juices could not be separated (Figure 2.2-22 a). On the other hand, the orange juice samples were mixed with the apple juice samples on the plot when only the electronic tongue data were considered (Figure 2.2-22 b). Using PLS modeling it is possible to classify correctly all the juice samples using both data sets simultaneously (Figures 2.2-22 c and 2.2-23). Therefore, the classification properties are improved when the combined data from both devices are used.

Wine flavor was studied using sensory fusion of taste and odor sensors [95]. The odor-sensing array consisted of four conducting polymer sensors. Measurements with two sensor systems were performed on four wine samples: white Esti di Montefiascone (Italy), white Chablis (France), red Bon Marché Mercian (Japan) and red Rosso di Montalcino (Italy). After data processing by PCA, all the wine samples were discriminated.

The combination of the potentiometric electronic tongue and electronic nose was applied to urine and milk analysis [85]. An electronic nose consisting of eight quartz microbalance sensors coated with different metalloporphyrins was developed at the University of Rome 'Tor Vergata'. Measurements were made on six samples of UHT and pasteurized milk, both fresh and sour. In medical experiments, seven samples of urine from healthy subjects and patients affected by kid-

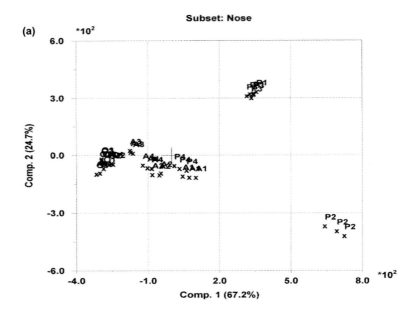

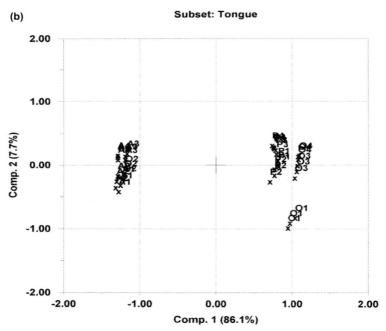

Figure 2.2-22. Score plots of juices from values obtained from (a) electronic nose, (b) tongue

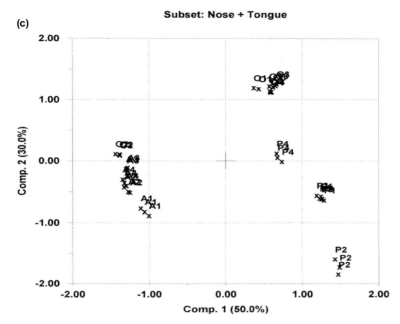

Figure 2.2-22 c. Both nose and tongue. O, Orange juice; A, apple juice; P, pineapple juice. Reprinted from Winquist, F., Lundstrom, I., Wide, P., The combination of an electronic tongue and an electronic nose, *Sens. Actuators B* **58** (1999) 512–517, with permission from Elsevier Science.

ney disease were measured with both systems. Urine samples were analyzed in the hospital laboratory for pH, specific gravity, protein, ketones, blood presence, etc. It was found that each system was able to distinguish different samples alone in both experiments, but the data fusion could improve classification.

Thus, the combination of an electronic tongue with an electronic nose ensures a higher recognition power of the combined system. This feature may be of significant practical impact when very similar samples are to be analyzed and reliable identification is crucial.

2.2.4.6 Environmental and Industrial Analysis

The main application of the electronic tongue is the analysis of foodstuffs, which is also popular owing to the analogy with biological sensing systems. Nevertheless, the sensors used in electronic tongues display sensitivity to many ions and other substances and can also be applied to their determination in inedible media. On the other hand, there is great concern in society about pollution of the environment by industrial discharges, which makes the determination of pollution of water an important task. Currently, this kind of analysis is usually performed

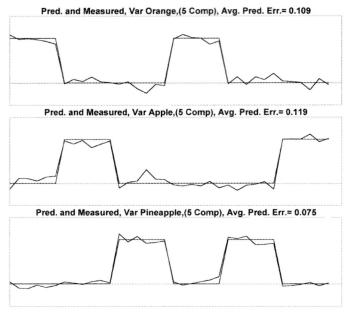

Figure 2.2-23. Predicted and measured values for a PLS model for the combination of an electronic nose and an electronic tongue. The thin line represents the true values, the thick line represents predicted values. Reprinted from F. Winquist, I. Lundstrom, P. Wide, *Sensors and Actuators B*, Vol. **58** (1999) 512–517, with permission from Elsevier Science.

in the laboratory and after sampling and is expensive. Therefore, a simple, rapid, and low-cost method for automated on-line evaluation of the content of many components would be of high value. Sensor systems and the electronic tongue appear to conform well to these requirements and thus represent a promising technique for environmental monitoring.

The potentiometric electronic tongue was applied to the analysis of modeled polluted ground waters, containing Cu^{2+}, Fe(III), Mn^{2+}, Zn^{2+}, Ca^{2+}, Mg^{2+}, Na^+, Cl^-, and SO_4^{2-} [53–55], and modeled flood waters from abandoned uranium mines, containing Fe(II), Fe(III), U(VI), and U(IV) [57]. The errors of determination were from 5 to 30%, which is acceptable for environmental monitoring especially considering the low concentration level of the components, which was about 10^{-7}–10^{-6} mol/L for metals.

A flow injection multisensor system was designed and applied to the determination of heavy metals and acid oxides in the flue gases from an incinerator [56]. The analysis was performed on solutions which were obtained by bubbling the flue gas through absorbing liquids. After the absorption, heavy metals were present in solution in ionic form, while acid oxides formed inorganic anions (NO_3^-, NO_2^-, SO_4^{2-}). The sensor array included up to 11 sensors. System calibration was performed in model solutions. Subsequently, samples from waste incineration plants were analyzed. The same samples were analyzed by inductively coupled plasma mass spectrometry and liquid chromatography with the aim of evaluating the performance of the multisensor flow injection system. The system was able to determine copper, lead, and cadmium with average errors of 10–20% and chloride, sulfate, nitrite, and nitrate with average errors of 5–15%.

The possibility of monitoring water contamination from factory drains using the taste sensor was investigated [96]. Substances such as cyanide, copper, and iron(III) could be determined in water using the device.

2.2.4.7 Biomedical Analysis

Medicinal analysis is another field where rapid and low-cost analytical methods are needed. Nowadays, ISEs are widely used in commercial blood analyzers. However, insufficient selectivity in multicomponent media hinders the further application of chemical sensors in this field. Another problem is the requirement for high accuracy of analysis, which should usually be within 1–2%, because the concentrations of many components in body fluids change only in a very narrow range (8–10%). The potentiometric electronic tongue, consisting of sensors with PVC plasticized membranes, was applied to component determination in dialysis solutions of an artificial kidney [58]. Owing to a specially elaborated measuring procedure, components of dialysis solutions such as Ca^{2+}, HCO_3^-, Na^+, K^+, and Cl^- and pH could be determined with errors as low as 2–4%, which complies with the requirements of this analysis. The possibility of determining the concentration of magnesium in the presence of calcium and the content of hydrophosphates using the electronic tongue was discovered. It was shown that the application of sensor arrays allows enhanced selectivity and decreased determination errors. Thus, electronic tongues could represent a new direction of chemical sensor application in medicine.

2.2.5 Conclusion

Electronic tongues are an emerging and prospective field of chemical sensor science. Such systems seem to be very useful as a quality control tool in the food industry and medicine and for environmental measurements, etc., in addition to being a multicomponent quantitative analysis instrument. A correlation between the electronic tongue output and human sensory perception has been found for many analytes. This is an exciting feature, promising novel, unusual, and practically important applications. Although some electronic tongues are already offered commercially, this is still an emerging scientific direction and the knowledge in this area still remains significantly empirical. Serious efforts are demanded in sensing mechanism studies and theoretical considerations, in the development of new sensor compositions, and in the procedures for the application of electronic tongues to practical tasks. We can expect new achievements in all these fields in the immediate future.

2.2.6   References

[1]   Haber, F., Klemensiewicz, Z., *Z. Phys. Chem.* (Leipzig) **64** (1909) 385–392.
[2]   Camman, K., *Working with Ion-Selective Electrodes;* Berlin: Springer, 1979.
[3]   Vlasov, Yu., Bychkov, E., *Ion-Sel. Electrode Rev.* **9** (1987) 5–93.
[4]   Bakker, E., Buhlmann, P., Pretsch, E., *Chem. Rev.* **97** (1997) 3083–3132.
[5]   Buhlmann, P., Pretsch, E., Bakker, E., *Chem. Rev.* **98** (1998) 1593–1687.
[6]   Buerk, D.G., *Biosensors: Theory and Application*; Lancaster: Technomic, 1993.
[7]   Janata, J., Josowicz, M., Vanysek, P., DeVaney, D. M., *Anal. Chem.* **70** (1998) 179R–208R.
[8]   Otto, M., Thomas, J.D.R., *Anal. Chem.* **57** (1985) 2647–2651.
[9]   Beebe, K., Kowalski, B., *Anal. Chem.* **60** (1988) 2273–2276.
[10] Beebe, K., Uerz, D., Sandifer, J., Kowalski, B., *Anal. Chem.* **60** (1988) 66–71.
[11] Forster, R.J., Regan, F., Diamond, D., *Anal. Chem.* **63** (1991) 876.
[12] Forster, R.J., Diamond, D., *Anal. Chem.* **64** (1992) 1721.
[13] Diamond, D., Forster, R.J., *Anal. Chim. Acta* **276** (1993) 75.
[14] Hartnett, M., Diamond, D., Barker, P.G., *Analyst* **118** (1993) 347–354.
[15] Van der Linden, W.E., Bos, M., Bos, A., *Anal. Proc.* **26** (1989) 329–331.
[16] Bos, M., Bos, A., van der Linden, W.E., *Anal. Chim. Acta* **233** (1995) 31–39.
[17] Gardner, J.W., Bartlett, P.N. (eds.), *Sensors and Sensory Systems for an Electronic Nose,* NATO ASI Series E: Applied Sciences; Dordrecht: Kluwer, 1992, Vol. 212.
[18] Gopel, W., Ziegler, Ch., Breer, H., Schild, D., Apfelbach, R., Joerges, J., Malaka, R., *Biosens. Bioelectron.* **13** (1998) 479–493.
[19] Stewart, G.F., Amerine, M.A., *Introduction to Food Science and Technology*; New York: Academic Press, 1973.
[20] Kawamura, Y., Kare, M.R., *Umami: a Basic Taste*; New York: Marcel Dekker, 1987.
[21] Moncrieff, R.W., *J. Appl. Physiol.* **16** (1961) 742–748.
[22] Persaud, K., Dodd, G.H., *Nature* **299** (1982) 352–355.
[23] Dickinson, T.A., White, J., Kauer, J.S., Walt, D.R., *TIBTECH* **16** (1998) 250–258.
[24] Nagele, M., Bakker, E., Pretsch, E., *Anal. Chem.* **71** (1999) 1041–1048.
[25] Legin, A., Vlasov, Yu., Rudnitskaya, A., Bychkov, E., *Sens. Actuators B* **34** (1996) 456–461.
[26] Vlasov, Yu., Legin, A., Rudnitskaya, A., *Sens. Actuators B* **44** (1997) 532–537.
[27] Legin, A., Rudnitskaya, A., Smirnova, A., Lvova, L., Vlasov, Yu., *J. Appl. Chem. (Russ.)* **72** (1999) 114–120.
[28] *Neural Computing*; Pittsburgh, PA: Neural Ware, 1997.
[29] Esbensen, K., Schonkopf, S., Midtgaard, T., Guyot, D., *Multivariate Analysis in Practice;* Norway: Camo ASA, 3rd edn., 1998.
[30] Martens, H., Naes, T., *Multivariate Calibration*; New York: Wiley, 1989.
[31] Toko, K., Hayashi, K., Yamanaka, M., Yamafuji, K., in: *Technical Digest of 9th Sensor Symposium*; 1990, pp. 193–196.
[32] Hayashi, K., Yamanaka, M., Toko, K., Yamafuji, K., *Sens. Actuators B* **2** (1990) 205–213.
[33] Toko, K., *Mater. Sci. Eng. C* **4** (1996) 69–82.
[34] Toko, K., *Meas. Sci. Technol.* **9** (1998) 1919–1936.
[35] Toko, K., *Sens. Actuators B* **64** (2000) 205–215.
[36] Oohira, K., Toko, K., Akiyama, H., Yoshihara, H., Yamafuji, K., *J. Phys. Soc. Jpn.* **64** (1995) 3554–3561.

[37] Hayashi, K., Toko, K., Yamanaka, M., Yoshihara, H., Yamafuji, K., Ikezaki, H., Toukubo, R., Sato, K., *Sens. Actuators B*, **23** (1995) 55–61.

[38] Iiyama, S., Suzuki, Y., Ezaki, S., Arikawa, Y., Toko, K., *Mater. Sci. Eng. C* **4** (1996) 45–49.

[39] Iiyama, S.; Toko, K.; Matsuno, T.; Yamafuji, K., *Chem. Sens.* **19** (1994) 87–96.

[40] Iiyama, S., Ezaki, S., Toko, K., Matsuno, T., Yamafuji, K., *Sens. Actuators B* **24** (1995) 75–79.

[41] Kikkawa, Y., Toko, K., Matsuno, T., Yamafuji, K., *Jpn. J. Appl. Phys.* **32** (1993) 5731–5736.

[42] Toko, K., Fukusaka, T., *Sens. Mater.* **9** (1997) 171–176.

[43] Nagamori, T., Toko, K., in: *Digest of Technical Papers of Transducers '99, Sendai, Japan*; 1999, pp. 62–65.

[44] Toko, K., Nagamori, T., *Trans. IEE Jpn. E* **119** (1999) 528–531.

[45] Akiyama, H., Toko, K., Yamafuji, K., *Jpn. J. Appl. Phys. Part 1* **35** (1996) 5516–5521.

[46] Toko, K., Akiyama, H., Chishaki, K., Ezaki, S., Iyota, T., Yamafuji, K., *Sens. Mater.* **9** (1997) 321–329.

[47] Toko, K., Yasuda, R., Ezaki, S., Fujiyishi, T., *Trans. IEE Japan E* **118** (1998) 1–5.

[48] Sasaki, Y., Kanai, Y., Uchida, H., Katsube, T., *Sens. Actuators B* **25** (1995) 819–822.

[49] Kanai, Y., Shimizu, M., Uchida, H., Nakahara, H., Zhou, C.G., Maekawa, H., Katsube, T., *Sens. Actuators B* **20** (1994) 175–179.

[50] Vlasov, Yu., Legin, A., *Fresenius' J. Anal. Chem.* **361** (1998) 255–260.

[51] Vlasov, Yu., Legin, A., Rudnitskaya, A., *Fresenius' J. Anal. Chem.* (2000) in press.

[52] Schoning, M., Schmidt, C., Schubert, J., Zander, W., Mesters, S., Kordos, P., Luth, H., Legin, A., Seleznev, B., Vlasov, Yu., *Sens. Actuators B* **68** (2000) 254–259.

[53] Di Natale, C., Davide, F., Brunink, J.A.J., D'Amico, A., Vlasov, Yu., Legin, A., Rudnitskaya, A., *Sens. Actuators B* **34** (1996) 539–542.

[54] Di Natale, C., Macagnano, A., Davide, F., D'Amico, A., Legin, A., Vlasov, Yu., Rudnitskaya, A., Seleznev, B., *Sens. Actuators B* **44** (1997) 423–428.

[55] Vlasov, Yu., Legin, A., Rudnitskaya, A., Butgenbach, S., Ehlert, A., *J. Appl. Chem. (Russ.)* **71** (1998) 1577–1580.

[56] Mortensen, J., Legin, A., Ipatov, A., Rudnitskaya, A., Vlasov, Yu., Hjuler, K., *Anal. Chim. Acta* **403** (2000) 273–277.

[57] Legin, A., Seleznev, B., Rudnitskaya, A., Vlasov, Yu., Tverdokhlebov, S., Mack, B., Abraham, B., Arnold, T., Baraniak, L., Nitsche, H., *Czech. J. Phys.* **49** (S1) (1999) 679–685.

[58] Legin, A., Smirnova, A., Rudnitskaya, A., Lvova, L., Vlasov, Yu., *Anal. Chim. Acta* **385** (1999) 131–135.

[59] Legin, A., Rudnitskaya, A., Seleznev, B., Vlasov, Yu., in: *Proceedings of 7th International Symposium on Olfaction and Electronic Noses, ISOEN, Brighton, UK*; 2000, p. 195.

[60] Legin, A., Rudnitskaya, A., Vlasov, Yu., Di Natale, C., D'Amico, A., *Sens. Actuators B* **58** (1999) 464–468.

[61] Legin, A., Rudnitskaya, A., Vlasov, Yu., D'Amico, A., Di Natale, C., in: *Proceedings of 8th International Conference on Electroanalysis, 11–15 June 2000, Bonn, Germany*; 2000, p. A27.

[62] Winquist, F., Wide, P., Lundstrom, I., *Anal. Chim. Acta* **357** (1997) 21–31.

[63] Winquist, F., Krantz-Rulcker, C., Wide, P., Lundstrom, I., *Meas. Sci. Technol.* **9** (1998) 1937–1946.

[64] Winquist, F., Holmin, S., Krantz-Rulcker, C., Wide, P., Lundstrom, I., *Anal. Chim. Acta* **406** (2000) 147–157.
[65] Wide, P., Winquist, F., Bergsten, P., Petriu, E.M., *IEEE Trans. Instrum. Meas.* **47** (1998) 1072–1077.
[66] Winquist, F., Lundstrom, I., Wide, P., *Sens. Actuators B* **58** (1999) 512–517.
[67] Lavigne, J.J., Savoy, S., Clevenger, M.B., Ritchie, J.E., McDoniel, B., Yoo, S.J., Anslyn, E.V., McDevitt, J.T., Shear, J.B., Neikirk, D., *J. Am. Chem. Soc.* **120** (1998) 6429–6430.
[68] Kondoh, J., Shiokawa, S., *Jpn. J. Appl. Phys., Part 1* **33** (1994) 3095–3099.
[69] Toko, K., Matsuno, T., Yamafuji, K., Hayashi, K., Ikezaki, H., Sato, K., Toukubo, R., Kawarai, S., *Biosens. Bioelectron.* **9** (1994) 359–364.
[70] Takagi, S., Toko, K., Wada, K., Yamada, H., Toyoshima, K., *J. Pharm. Sci.* **87** (1998) 552–555.
[71] Tagaki, S., Toko, K., Wada, K., Ohki, T., in: *Digest of Technical Papers of Transducers '99, Sendai, Japan;* 1999, pp. 1638–1641.
[72] Iiyama, S., Iida, Y., Toko, K., *Sens. Mater.* **10** (1998) 475–485.
[73] Nagamori, T., Toko, K., Kikkawa, Y., Watanabe, T., Endou, K., *Sens. Mater.* **11** (1999) 469–477.
[74] Legin, A., Rudnitskaya, A., Vlasov, Yu., Di Natale, C., Mantini, A., Mazzone, E., Bearzotti, A., D'Amico, A., *Alta Frequenza* **10** (1999) 1–3.
[75] Legin, A., Rudnitskaya, A., Vlasov, Yu., Di Natale, C., Mazzone, E., D'Amico, A., *Electroanalysis* **11** (1999) 814–820.
[76] Legin, A., Rudnitskaya, A., Vlasov, Yu., Di Natale, C., Mazzone, E., D'Amico, A., *Sens. Actuators B* **65** (2000) 232–234.
[77] Legin, A., Rudnitskaya, A., Seleznev, B., Vlasov, Yu., *Food Sci. Technol.* (2001) in press.
[78] Iiyama, S., Yahiro, M., Toko, K., *Sens. Mater.* **7** (1995) 191–201.
[79] Fukunaga, T., Toko, K., Mori, S., Nakabayashi, Y., Kanda, M., *Sens. Mater.* **8** (1996) 47–56.
[80] Legin, A., Rudnitskaya, A., Vlasov, Yu., Di Natale, C., Davide, F., D'Amico, A., *Sens. Actuators B* **44** (1997) 291–296.
[81] Legin, A., Rudnitskaya, A., Vlasov, Yu., in: *Proceedings of 6th International Symposium Kinetics in Analytical Chemistry, KAC '98, Kassandra Halkidiki, Greece, 16–19 September, 1998;* 1999, p. 106.
[82] Rudnitskaya, A., Legin, A., Salles, C., Mielle, P., in: *Proceedings of 8th International Symposium on Olfaction and Electronic Noses, ISOEN, 25–28 March 2001, Washington, DC.*
[83] Toko, K., Iyota, T., Mizota, Y., Matsuno, T., Yoshioka, T., Doi, T., Iiyama, S., Kato, T., Yamafuji, K., Watanabe, R., *Jpn. J. Appl. Phys., Part 1* **34** (1995) 6287–6291.
[84] Yamada, H., Mizota, Y., Toko, K., Doi, T., *Mater. Sci. Eng. C,* **5** (1997) 41–45.
[85] Di Natale, C., Paolesse, R., Macagnano, A., Mantini, A., D'Amico, A., Legin, A., Lvova, L., Rudnitskaya, A., Vlasov, Yu., *Sens. Actuators B* **64** (2000) 15–21.
[86] Legin, A., Rudnitskaya, A., Lvova, L., Vlasov, Yu., D'Amico, A., Di Natale, C., Paolesse, R., in: *Proceedings of 5th Italian Conference on Sensors and Microsystems, Lecce, Italy;* Singapore: World Scientific, 2000, pp. 263–269.
[87] Legin, A., Rudnitskaya, A., Vlasov, Yu., Di Natale, C., Mazzone, E. D'Amico, A., in: *Proceedings of International Conference Eurosensors XIII, The Hague, The Netherlands;* 1999, pp. 1017–1020.
[88] Toko, K., Murata, T., Matsuno, T., Kikkawa, Y., Yamafuji, K., *Sens. Mater.* **4** (1992) 145–151.

[89] Arikawa, Y., Toko, K., Ikezaki, H., Shinha, Y., Ito, T., Oguri, I., Baba, S., *Sens. Mater.* **7** (1995) 261–270.

[90] Arikawa, Y., Toko, K., Ikezaki, H., Shinha, Y., Ito, T., Oguri, I., Baba, S., *J. Ferment. Bioeng.* **82** (1996) 371–376.

[91] Vlasov, Yu., Seleznev, B., Ivanov, A., Rudnitskaya, A., Legin, A., in: *Proceedings of 5th Italian Conference on Sensors and Microsystems, Lecce, Italy;* Singapore: World Scientific, 2000.

[92] Kikkawa, Y., Toko, K., Yamafuji, K., *Sens. Mater.* **5** (1993) 83–90.

[93] Imamura, T., Toko, K., Yanagisawa, S., Kume, T., *Sens. Actuators B* **37** (1996) 179–185.

[94] Legin, A., Rudnitskaya, A., Seleznev, B., Vlasov, Yu., Velikzhanin, V., in: *Proceedings of 8th International Symposium on Olfaction and Electronic Noses, ISOEN, 25–28 March 2001, Washington DC.*

[95] Baldacci, S., Matsuno, T., Toko, K., Stella, R., De Rossi, D., *Sens. Mater.* **10** (1998) 185–200.

[96] Taniguchi, A., Naito, Y., Maeda, N., Sato, Y., Ikezaki, H., *Sens. Mater.* **11** (1999) 437–446.

List of Symbols and Abbreviations

Symbol	Designation
a_i	activity of primary ion
E	potential difference (e.m.f.)
E°	standard potential
F	Faraday constant
F	nonselectivity factor
i	primary ion
j	interfering ion
K	average signal-to-noise ratio
$K_{i,j}$	selectivity coefficient
n	number of components
R	gas constant
s	standard deviation
S	response slope
T	absolute temperature
z_i	electric charge of primary ion
z_j	electric charge of interfering ion
τ	strength of bitterness

Abbreviation	Explanation
ANN	artificial neural network
DA	decyl alcohol
DOP	dioctyl phosphate
DOPP	dioctyl phenylphosphonate
FET	field effect transistor
ISE	ion-selective electrode
LAPS	light addressable potentiometric sensor
LAPV	large amplitude pulse voltammetry
MOSFET	metal oxide semiconductor field effect transistor
OA	oleic acid
OAm	oleylamine
PCA	principal component analysis
PEG	poly(ethylene glycol)
PLD	pulse laser desorption
PLS	partial least square regression
PS	polystyrene
PVC	poly(vinyl chloride)
SAPV	small amplitude pulse voltammetry
SH-SAW	shear horizontal surface acoustic wave
SPV	surface photovoltage
TOMA	trioctylmethylammonium chloride
UHT	ultra-high temperature
WPNI	whey-protein nitrogen index

2.3 Understanding Chemical Sensors and Chemical Sensor Arrays (Electronic Noses): Past, Present, and Future

J. R. STETTER and W. R. PENROSE, Illinois Institute of Technology, Chicago, USA

Abstract

The use of sensor arrays, embodied in electronic noses, to characterize complex samples is an active and dynamic expression of research in chemical sensors. In this review, we emphasize sensors, instrumentation, and applications aspects of electronic nose technology. In addition, we have added to the existing historical description of electronic nose development, projected likely directions into the future, and evaluated research needs that are a prerequisite to eventual success of electronic nose technology.

Keywords: Sensor arrays; electronic nose; artificial olfaction; analytical chemistry

Contents

Dedicated to the memory of our good friend and colleague Wolfgang Göpel, who died in June 1999 in the prime of his career. He dedicated part of his precious life to sensors, chemical sensor arrays and electronic noses.

Intelligence, intensity, depth, vision,
Marked your science call.
Passion, grit, and oft unappreciated wit,
Brought new knowledge to all.

2.3.1   Introduction

The field of chemical sensors has expanded dramatically in the past twenty-some years. The first International Meeting on Chemical Sensors (IMCS) in 1983 in Fukuoka, Japan [1], was the one at which I first met Wolfgang Göpel. He had just returned to Germany to head the Institute for Physical and Theoretical Chemistry in Tübingen.

The IMCS has been held eight times, as has the 'Transducers' series of conferences, since 1983. These conferences were the result of the international interest in the topic generated at an earlier Material Research Society technical session in 1980. Recent conferences on chemical sensors can be found on the websites of many technical organizations [2].

A major enthusiast and a leader in the explosive growth of the chemical sensor field was the late Wolfgang Göpel (1945–1999), scientist, friend and colleague. His efforts encompassed the tireless pursuit of better chemical and physical sensors as well as a more complete understanding of the science behind a broad range of sensors. It is in this tradition of sensor science that this perspective and update is presented, with special focus on gas sensors and the special properties of sensor arrays, including those applications often known as 'electronic noses'. We will describe current capabilities and some recent publications in a historical context in an attempt to make useful comments about future capabilities and trends. This review will discuss sensor arrays or electronic noses from the point of view of the sensors, analytical chemistry, and recent applications including the successes of artificial olfaction. Also, we will provide interpretive comments concerning the origin of unique sensing capabilities and the debunking of popular myths that have arisen about the electronic nose technology. The biology of olfaction and pattern recognition has been covered in depth by two recent, excellent reviews, and will not be covered in any depth here. The focus of this review is to summarize current understanding in new and meaningful ways that are complementary and not redundant to earlier issues of *Sensors Update*.

The scientific world of chemical sensors includes those that operate in gases, liquids, and solids. While arrays of gas sensors are known as 'electronic noses', arrays of liquid sensors are becoming known as 'electronic tongues'. Sensing in each phase presents different technical challenges, but chemical sensors in all phases share some common characteristics. The following discussion centers mainly around gas sensors, but will often apply to all classes of chemical sensors. Although a discussion of sensor arrays is the main purpose of this paper, a discussion of individual sensors with a focus on those most frequently found in arrays is presented. The understanding of sensor principles and metrics is essential to the understanding of the performance of the resulting sensor arrays.

Chemistry, quality control, and process applications in the food industry have so far been the most common applications of sensor array technology. Many foods give strong signals on chemical sensors, compared with other sample types. Moreover, the great complexity of odors and flavors, as well as their subjective nature, frequently defeats conventional analytical methods. The potential for rapid analysis at lower cost has also made electronic nose technology attractive to the food industry. In our laboratory, coffee, tea, cooking oils, cheeses, and beer have been used for student projects because they provide fine distinctions and good sensor responses [3].

2.3.2   The Literature on Chemical Sensor Arrays and Electronic Noses

The literature on chemical sensors, sensor arrays, and electronic noses has been regularly reviewed in *Sensors Update*. There has been a steady and rapid increase in the rate of appearance of references to electronic noses. From January

1994 to June 2001, there are over 360 returns on the keywords 'electronic nose(s)', 'artificial olfaction', or 'sensor array(s)' from the *Current Contents* database ('sensor array' returns were filtered to exclude optical imaging and similar physical sensor arrays). A crude classification of these terms is shown in Table 2.3-1. Many of these have been selected for inclusion in the reference list for this chapter.

Not surprisingly, much of the effort over the past 8 years has been in the selection of gas sensors and instrument design. The novelty of the field has attracted a burst of development. Since sensor array instruments are not difficult to build, it is not surprising that many alternative designs and sensor selections are published each year.

Major reviews on sensor arrays and electronic noses have appeared in the present series. Volume 2 of *Sensors Update* contained a review emphasizing pattern recognition [4], while a complementary article in Volume 3 addressed primarily the biology of olfaction [5]. We will not revisit these subjects, but will emphasize progress in sensors, sensor arrays, and electronic noses since the 1991 review by Vaihinger and Göpel [6] in the foundation volumes of this series.

Other reviews have appeared from time to time, with increasing frequency. As we have discovered ourselves in the preparation of this review, the field is rapidly becoming too large to cover comprehensively in a single contribution, and so most recent reviews have emphasized one or another aspect of the technology. In a recent *Chemical Reviews*, the properties of sensor arrays were covered by Albert et al. [7], while Jurs et al. [8] surveyed the literature on pattern classification algorithms. General reviews, that might be used as a first introduction to the field, are already becoming quite dated, although still useful. We would suggest

Table 2.3-1. Classification of primary and review papers citing the terms 'electronic nose(s)', 'artificial olfaction', or 'sensor array(s)' for the time periods 1994–1999 and January 2000–June 2001. Classification was by inspection of titles and keywords. Since many different types of physical property sensors, such as imaging devices, are referenced under 'sensor arrays', only those papers referring to chemical sensors are included in this listing

Category	Number of papers	
	1994–1999	2000–2001 (18 months)
Sensors and hardware design	50	46
Pattern classification and theory	25	16
Applications – food, agriculture	65	36
Applications – medical	6	9
Applications – bacteriology	5	8
Relationship to biological olfaction	5	6
General, reviews, miscellaneous	50	40
Total	206	161

the papers by Dickinson et al. [9] and Nagle et al. [10]. A textbook entitled *Electronic Noses* was recently published by Gardner and Bartlett [11], although it emphasizes chemical sensors with only relatively brief mention of sensor arrays and electronic noses. Excellent explanations and expositions on the electronic nose and sensor arrays can be reached at several academic institutions through the NOSE website [12].

The term 'electronic nose' came into more common use after it appeared in the title of a major monograph on the electronic nose, edited by Gardner and Bartlett [13], which resulted from a 1991 NATO-sponsored meeting convened in Reykjavik, Iceland. To avoid the pejorative connotation of the term 'electronic nose', some authors prefer the term 'artificial olfaction', while others prefer the more formal (if ambiguous) 'chemical sensor arrays'. Another multi-authored review, edited by Kress-Rogers [14], emphasized biomedical applications of electronic noses, along with biosensors.

Electronic noses have been the subject of at least one symposium series. A loose organization of researchers, first assembled by the French company Alpha MOS, has sponsored an annual International Symposium on Olfaction and the Electronic Nose. Even though the conference no longer has an institutional base, it held its eighth consecutive symposium in March 2001 in Washington, DC, under the aegis of the Electrochemical Society [15]. Some of the more recent meetings of this group have also resulted in published proceedings volumes [16]. The next ISOEN meeting is scheduled for September 2002 in Rome, Italy.

Because the electronic nose paradigm has not yet been sanctified by the passage of long periods of time, considerable ingenuity continues to be invested in exploration of unique designs. Walt's group at Tufts University (Amherst, MA) has explored the optical behavior of polymer beads with different surface chemistry, coated with a solvatochromic fluorescent dye [17]. A large number of different bead types are mixed together. They are attached to the ends of optical fibers, where each may be individually interrogated with a light pulse. Because of the method in which the beads and fibers are assembled, the assignment of 'sensors' (the sensitive polymer beads) to data channels (individual fibers) is random. A neural network is used to create an association between each sample type and the pattern of responses. In a real sense, this device resembles the structure and ontogeny of the mammalian olfactory sense more than many other realizations of electronic nose technology. The approach has been given the trade name of BeadArray and is being commercialized by Illumina Inc. (San Diego, CA). A unique hint at the future can be gained by examination of the result of an experiment that would place several of these optical arrays inside a physical model of a dog's nose. This experiment would produce data that consisted of arrays of array data separated in space and time with some chemical filtration (ie, changes) in between. This is very much like the mammalian olfactory system and should produce enhanced results with the additional information content.

Another unique variation on the electronic nose is the colorimetric sensor array recently proposed by Rakow and Suslick [18] for identifying solvent vapors. These authors noted that most odorous compounds have at least some Lewis base activity, and would bind to the central atom of a tetraphenylmetalloporphyrin, changing its color in a unique way. The authors spot a series of different metalloporphyrins on to a silica substrate in a stereotyped pattern and expose it to various solvent vapors. Each vapor produced a unique pattern of colors. Such an 'electronic nose' is not necessarily electronic, and does not even require an instrument! The authors, however, have proposed a CCD camera to reduce the array colors to digital form for processing by automatic means.

2.3.3   Chemical Sensors as Related to Analytical Chemistry and Arrays

To understand chemical sensor arrays, one should first understand chemical sensors and how their information is developed. The specific properties of each chemical sensor used in the array provide the chemical dimensionality to the array data and hence determine its sensing capability. Just as different human noses are different among people and their ages, sensors are also differently selective and sensitive. Some sensors respond to volatile organics and others to permanent gases. The wide range of chemical sensors now available is at least in part the reason for the evolution of arrays and electronic noses and, more recently, electronic tongues.

For the remaining sections, we will define a 'class' of chemical sensors as one operating on a common transducer platform. Sensors using the DC resistance of a heated metal oxide semiconductor (whether 'Taguchi' SnO_2 sensors or higher temperature Ga_2O_3 sensors) would then be members of the same class, electrochemical sensors, and the subclass of impedance-based sensors. Sensors within a class with distinct properties can be referred to as 'types.' Thus, the differently doped SnO_2 sensors, made to have different selectivities for hydrogen, methane, and carbon monoxide, are 'types' within the 'class' of electrochemical-impedance sensors and are often called heated metal oxide semiconductor or MOX sensors. While the nomenclature of 'classes' and 'types' is not standard or established in the literature yet, we will use it here to add clarity to the discussion of sensors and arrays.

Further, for the sake of this discussion, we can define four distinct kinds of problems each demanding a different analytical perspective as described in Table 2.3-2. Of course, any instrument can be used for any problem it solves. Infrared (IR) spectrometers have been used for the relatively simple online CO measurement for a long time. But the above comments point to typical analytical devices and instruments used to solve the analytical problem at hand. While it has been the goal for many years to obtain a sensor to detect trace explosives like TNT (a type A problem), it would be difficult to imagine a single sensor capable of complex mixture analysis. Even high-resolution IR spectrometers require additional help, including experienced human operators, in order to arrive at useful answer.

There are basically three types of 'sensors' or tiny self-contained analytical systems that can perform chemical analysis:

Table 2.3-2. Four generic kinds of analytical problems

Type A: A single component in a complex matrix, eg, CO in air or Fe in iron ore. The matrix (that portion of the sample that is not the target of the analysis) may be consistent or highly variable. A special subset of this class is trace analysis – the needle in the haystack problem. This is the class of analysis that single sensors often target. The carbon monoxide sensor is one of the two or three most common chemical sensors in commerce

Type B: The major components of a mixture. Air consists of five important components, nitrogen, oxygen, carbon dioxide, water vapor, and argon. Such analysis can be carried out by gas chromatography or in part by infrared analysis or by oxygen and relative humidity sensors

Type C: Complex mixtures. There are 640 characterized components in the headspace gas above coffee. Extensive sample preparation and complex instrumentation and training is generally required for this type of task

Type D: Determination of a subjective, collective, or arbitrary endpoint. Examples of such endpoints include lower explosive limits or lower flammable limits, toxic concentrations, and assessments of quality of a raw material such as grain, or of a finished product such as a food for sale. One may not be able to provide a specific and complete molecular explanation of the endpoint or the endpoint may not refer to a specific molecule or set of molecules and their concentrations. This class of analytical problem is addressed directly by the human or electronic nose and/or tongue as well as a select few sensors (eg, the combustible gas sensor)

Type 1: Chemical Sensors and Biosensors that measure quantity or quality of a substance because of its chemical reaction with the sensor interface, while the extent of this reaction is converted (transduced) to a measurable electrical signal by communication of the interface with a physical transducer.

Type 2: Microinstruments that measure molecular properties, like small IR spectrometers. These are typically physical sensors that measure a molecular property. Chemists call this molecular spectroscopy and it can be done with electromagnetic (NMR, IR, UV, x-ray), thermal (TGA, DTA, thermal conductivity), electronic (Auger), mechanical (sonic), or many other forms of energy interacting with the molecule. We extract molecular information from measuring the change in the physical energy that was caused when the energy interacted with the molecule. These sensors have been called physico-chemical sensors, but this distinction is superfluous. An example of such a sensor is the nondispersive IR sensor commonly used to measure carbon dioxide in ambient or exhaled air; some such commercial sensors are less than 25 cm^3 in size and the latest research has given us microfabricated versions [19].

Type 3: Micro-Total Analytical Systems (μ-TAS), as they have become known, that are small analytical systems incorporating sampling, separation, mixing, transporting, detecting, and all types of such analytical processes. They are generally made by silicon lithography, plastic molding, or imprinting methods [19].

While these definitions are somewhat arbitrary, they serve a useful purpose here so that we can discuss the field of sensors in an organized manner. As far as a customer or application is concerned, a definition of chemical sensor could be 'anything that goes into the small box and solves the analytical problem as long as the chemical analysis is correct.' But when we develop or discuss technology, the utility of the definition is proved by the increase in understanding and communication it provides. A more complex and comprehensive definition of a chemical sensor is given in Table 2.3-3.

The chemical sensor is 'a small self-contained integrated system of parts, that, as the result of a *chemical* interaction or process (ie, a reaction) between the analyte and the device, transforms chemical or biochemical information of a quantitative or qualitative type into an analytically useful signal.' This is a type 1 device and all three types are compared in Figure 2.3-1. Conceptually, every chemical sensor consists of two domains: the physical transducer and the chemical interface layer or receptor domain (Figure 2.3-1 a). At the chemical interface, the analyte interacts chemically with a surface or coating or catalyst that is part of the device and produces a change in physical/chemical properties. These changes are measured by the *transducer* domain, which monitors this change and generates a proportionally related electrical signal. Sometimes these domains are intermixed and the same, sometimes not. Sensors are frequently classified and named by their transduction method, eg, conductimetry, potentiometry, amperometry, gravimetry, which includes SAW (surface acoustic wave) and QMB (quartz crystal microbalance) transduction, optical (fiber optic, spectrometric, and refractometric), metal oxide semiconductor (MOX) (heated metal oxide chemiresistive), conductive polymer chemiresistive, polymer composite chemiresistive, or capacitive. Alternately, they may be named according to their structure, as in MOS (metal oxide semiconductor) or MIS

Table 2.3-3. Properties typically associated with chemical sensors

They all have:

- **a sensitive layer that is in chemical contact with the analyte gas**
- **a change in the chemistry of the sensitive layer**, ie, it is not the same after exposure as before exposure to the analyte because a stoichiometric change has occurred (a reaction)
- **no moving parts** in this mechanism not with standing mass transport in reactions

They all respond to the presence of a chemical with an electrical output and this means the sensitive layer is on a platform that allows transduction or coupling of the sensitive layer changes to a transducer that relates the change to electrical signals

They are physically small

They operate in real time, controlled by thermodynamic and kinetics of a chemical reaction; although the readout may be temporal (notwithstanding physical limitations due to design)

They do not necessarily measure a single or simple physical or chemical property

They are typically less expensive and more convenient than an equivalent instrument for the same chemical measurements

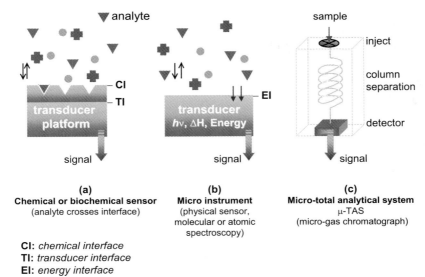

(a)
Chemical or biochemical sensor
(analyte crosses interface)

(b)
Micro instrument
(physical sensor,
molecular or atomic
spectroscopy)

(c)
Micro-total analytical system
μ-TAS
(micro-gas chromatograph)

CI: *chemical interface*
TI: *transducer interface*
EI: *energy interface*

Figure 2.3-1. Three categories of miniature analytical devices. (**a**) Chemical sensor, where a chemical reaction is linked to a transduction mechanism. (**b**) Microinstrument, in which an external field or energy source is used to measure nondestructively a physical property such as optical absorption. (**c**) Micro-total analytical system μ-TAS, which carries out a multistep analytical procedure, and generally incorporates a sensor or a microinstrument as the detection device.

(metal insulator semiconductor) sensors. A third name is derived from the chemical reaction such as 'sorption' sensors or 'catalytic' sensors.

The schematic of a chemical sensor (Figure 2.3-1 a) is compared with a microinstrument illustrated in Figure 2.3-1 b and a μ-TAS in Figure 2.3-1 c. Not included in the sensors here are the sampling systems or inlets, the housings, and the readout that is necessary to get an analytically meaningful result. These parts are often backward integrated into the sensor device as far as possible to meet simplicity, performance, or cost objectives. The physical sensor, Figure 2.3-1 b, sends out and receives a form of energy that interacts with the analyte molecule and, most distinctly, has no chemical interface that, per the above definition, is a required part of the chemical sensor. In the physical sensor for chemical analysis category are measurements of thermal conductivity (thermal sensor), infrared absorption (optical sensors), paramagnetism (usually for oxygen), NMR, etc. The analytical chemist calls this molecular spectroscopy when electromagnetic energy is involved and we could call this micro- or nano-molecular spectroscopy to differentiate it from the larger instrumentation found in the laboratory.

In the μ-TAS group illustrated in Figure 2.3-1 c, the portable gas chromatograph or 'lab on a chip' is illustrated. Also, we could include in this group the mass spectrometer on a chip because it contains an inlet, ionization chamber, accelerator, mass filter, and ion detector in a vacuum and represents a process for molecule or atom isolation and subsequent detection. This approach integrates as many analytical processes as are required to solve the problem at hand.

Note well that the chemical or biosensor has a reactive interface. Every chemical reaction is characterized by a equilibrium constant with its attendant thermodynamic characteristics that can be used to understand the sensor. Note also that the interface is changed during sensing, ie, it is stoichiometrically a different material before and after analyte exposure! This is in stark contrast to the physical sensor or microinstrument as defined above.

Figure 2.3-2 illustrates one way to represent the categories of chemical sensors arranged by their physical principle, ie, the physical sensor that transduces the signal. In this manner we see reactions at the interface that are catalytic or absorptive in all classes of sensors. For discussion purposes, it is also useful to organize chemical sensors by the type of chemical reaction that occurs at the interface. In Table 2.3-4 we illustrate this rarely taken approach in the sensor literature. Each reaction is characterized by a different type of chemical reaction, an equilibrium constant, and attendant kinetic parameters. These chemical parameters will determine to a large extent the observed sensor performance and provide an explanation for the observed sensor selectivity and sensitivity.

Each of these chemical reactions can be used on virtually any transducer platform. For example, there have been many proposals for sensor development where the proposer says that a sensor will be built for a gas, perhaps CO, on a platform, perhaps optical, and 'all that is needed' is the right coating to make the sensor work. The next logical step seems to be to test 1000 polymer coatings for this purpose, but never the proposed chemical reaction that will be used to achieve the selectivity and sensitivity. It is a simple matter to understand that the

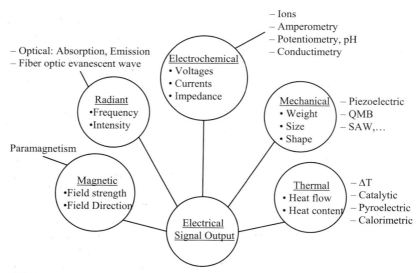

Figure 2.3-2. Organization of chemical sensors by class (transduction mechanism).

Table 2.3-4. Possible reactions at the chemical interface

Adsorption	$A[gas] + S[surface\ site] = AS[surface]$ K_{ads}
Physisorption	$V_{ads} = A\,e^{-dG/kT}$
Chemisorption	
Partitioning	$K = C_m/C_s$
Acid-base	$HA + KOH = H_2O + K^+ + A^-$ K_a or K_b
Precipitation	$Ag^+[aq] + NaCl[aq] = AgCl[s] + Na^+$ K_{sp}
Ion exchange	$H^+[aq] + Na^+[surface] = H^+[surface] + Na^+[aq]$ K_{iex}
Oxidation/reduction	$CO + \frac{1}{2}O_2 = CO_2$ K_{rxn}

polymer coating is not a very promising approach for CO, because it works by partitioning, an equilibrium reaction of the analyte between the mobile (or gas) and stationary (or liquid) phases. The amount of analyte that is absorbed into the stationary phase – the polymer in this case – is governed by this equilibrium constant. If no gas partitions into the surface coating, no change can be induced that will be detected by the platform. Typical gas chromatography (GC) phases attract gases by intermolecular or condensation forces, which are very weak in permanent gases like CO. This is the reason QMB and SAW devices that use a polymer coating have very low sensitivity to CO.

Logic would dictate the use of a chemical interface that was designed to take advantage of the special properties of CO. Since CO strongly adsorbs to certain metals and is oxidizable even at very low concentrations, the selection of a metal-based coating such as a metalloporphyrin, or an oxidation as the chemical interface is more promising. In fact, catalytic combustion on a metal oxide semiconductor, or an electrocatalytic oxidation, such as occurs in the amperometric CO sensor, are the favored reactions for common CO sensors. So consideration of the reaction type is extremely useful when trying to predict sensor characteristics while consideration of the transducer platform is less useful for this purpose.

When we categorize the chemical sensors by the chemical reaction of the chemical interface, it can make them easier to understand. Understanding makes it possible to avoid mistakes. Suppose, for example, we are searching for a sensor that will measure benzene in the presence of another aromatic hydrocarbon, such as naphthalene. Using a carbon sorbent operating with a physisorption reaction is not a promising approach. The intermolecular forces that are responsible for physisorption of aromatic compounds to carbon are going to be similar for all aromatics, and they will be stronger for polyaromatic compounds than for benzene. It is not reasonable to promise to develop a surface that will detect benzene to the exclusion of naphthalene or other aromatics using partition coefficients alone. The carbon coating will always adsorb more naphthalene than benzene at any temperature or pressure. This is a qualitative statement of the thermodynamics of gas adsorption, from which we can derive that physisorption forces are similar to condensation. There will be no high sensitivity physisorption unless the substrate is cooled to at or below the boiling point of the analyte. It also explains why, on sorption sensors, naphthalene is more 'sensitively' determined because at any given gas pressure, more naphthalene than benzene will be condensed. This is not to say that it is impossible to detect benzene selectively in the presence of naphthalene, but it must be done by some selective interaction beyond simple surface adsorption, perhaps by molecular size. Molecular sieves, polymer imprinting, or some specific chemical reaction are better choices for the choice of a detection mechanism.

Further, selectivity is only achieved if the reaction is selective among the sample constituents. That is, an acid coating on the surface will detect all volatile bases. The relative sensitivities to bases will be determined by the relative strength of the acid-base equilibrium constants.

Sensitivity is also a function of temperature. The temperature response of amperometric sensors is a function of the temperature behavior of several electrode processes with different parameters, which are exponential, but may oppose or enhance one another. Sensor response, therefore, may increase or decrease with temperature, but perhaps not in a predictable way. It is for such reasons that many amperometric sensors are designed with a pinhole to limit the analyte entry into the sensor. In this way, the response is limited by a single, predictable physical process such as diffusion, which varies with the square root of temperature.

In summary, to understand the origins of the performance characteristics like sensitivity, selectivity, response time, and stability, we examine the type of chemical reaction and its relevant thermodynamics and kinetics. Of course, this can be an oversimplification if the sensor device is limited by the transducer platform or physical housing. But when studying and reporting results with sensors, testing and experimentation should clearly address the limiting factors be they the chemical interface, the physical system, or the transducer.

With this guidance, one can identify and resolve some of the myths about sensors, and better direct the development of sensor systems for any given application. In the latter discussion, we illustrate these principles considering that all E-noses are not created equal. Some are biased toward the detection of vapors and some permanent gases, some operated over a broad range of concentrations, and others over a narrower window. It also becomes clear why heterogeneous arrays can perform many analytical tasks better than a homogeneous array. The heterogeneous array is capable of responding to more classes and types of compounds in the sample, over a broader range, and can produce a larger information output.

Finally, we summarize some of the observations that we can make about chemical sensors by collecting their common properties. Most often chemical sensors do not necessarily measure a simple physical property. A simple illustration is the response of a QMB polymer coating to a given analyte vapor, say benzene, at high and low relative humidity. Because of the presence of water vapor, we are no longer measuring the simple partitioning of benzene into the coating but the partitioning in coatings in different hydration states. Moreover, the interactions of water or matrix with the coating and benzene with the coating are interdependent. The coating may also change stiffness in response to analyte, water vapor, or other interfering compounds, which also affects the QMB response.

Piezoelectric gravimetric sensors respond not just to the mass loadings, but also to the shear modulus, or stiffness, of the coating. The final observed response will be the summed output of several reactions and several physical/chemical property changes that have occurred in the systems because of the interfacial reactions. These may or may not be linearly additive.

2.3.4 Sensor Arrays and Electronic Noses

2.3.4.1 Properties of Chemical Sensor Arrays and Electronic Noses

A chemical sensor array once required little definition. It was merely a collection of sensors, with the only restrictions being that the sensors were all exposed to the same or nearly the same sample, and the responses were interpreted together. Exercise of the imagination in developing new electronic nose modes and embellishments has since complicated the definition. Often, it is necessary to specify the order of the sensors in the array, because some sensors act upon the sample and change it. An example of this is MOX sensors, which efficiently oxidize the sample and change the response of other, downstream sensors. The sensors may be of the same or mixed classes, and each will exhibit the features, advantages, limitations, and problems of its respective class. These arrays are termed homogeneous and heterogeneous arrays, respectively. Some of the classes of chemical sensors that have been used in arrays are listed in Table 2.3-5.

An additional complication in defining a sensor array is the emergence of the term *virtual sensor.* A virtual sensor refers to a large number of distinct responses gathered from a single or smaller number of physical sensors. An example is the use of the mass spectrometer as a sensor, in which a single physical detector (a photomultiplier) is used to gather hundreds of distinct *m/ze* responses, any of which can be isolated and treated as a separate sensor response. Other examples of virtual sensors are sample fractions eluted from a gas chromatograph column, selected wavelengths from an infrared spectrometer, and the early use of programmed sample preprocessing to obtain multiple pattern elements from single electrochemical sensors in a sensor array [20, 21].

The important feature of the sensor array lies in the collected responses of its constituent sensors. We are in the habit of calling the ordered array of sensor responses a 'pattern', because at this stage of description we are beginning to enter the territory of pattern recognition or pattern classification, and we should properly adopt the terminology of this field. The patterns of sensor responses are uniquely determined by each sample. In the simplest interpretation, we can plot the sensor response patterns as histograms. For sufficiently different chemicals, these can usually be distinguished by eye (Figure 2.3-3).

The sensor array is one of three defining components of an 'electronic nose' (Figure 2.3-3). Although there is no universally accepted definition of an elec-

Table 2.3-5. Chemical sensors used in arrays and E-noses

Transduction mode (chemical interface)	Comments
Conductimetric (conductive polymer)	Earliest commercial E-nose (AromaScan); easily made by electrodeposition (moisture, temperature, history-sensitive)
Conductimetric (metal oxide semiconductor)	Used in largest number of E-noses (moisture and history-sensitive) WMA Airsense Electronic Nose Fox n000 (Alpha MOS) NST 3200 series (AppliedSensor was: Nordic Sensor, Mo-Tech FreshSense (Element Ltd.) MOSES (Lennartz Electronic)
Mechanical (sorptive polymer, quartz microbalance)	Better linearity than above sensors Libranose (University of Rome 'Tor Vergata') MOSES (Lennartz Electronic)
Mechanical (sorptive polymer, SAW sensor)	VaporLab (Microsensor Systems Inc.) The Electronic Nose: combined with short-column GC (Electronic Sensor Technology)
Electrochemical (amperometric gas sensor)	Linear sensor response allows normalization of patterns for concentration independence MOSES Electronic Nose (Lennartz Electronic) CPS-100 (Transducer Research Inc.)
Electrochemical, Potentiometric ion-selective sensor	Electronic tongue
Conductimetric Polymer composite sensors	Cyranose 320 (Cyrano Sciences Inc.)
Conductimetric, ChemFET	NST 3200 series (AppliedSensor was: Nordic Sensor, Mo-Tech)
Mass spectrometer	HP 4440A (Agilent Technologies) Prometheus, Kronos (Alpha MOS)
Optical, fluorescent sorptive beads on fiber optics	BeadArray (Illumina Inc., San Diego)

tronic nose, it is in all cases a gas sensor array packaged for use in practical analytical problems. The use of the term 'electronic nose' became more common after the monograph by Gardner and Barlett [13], although nowhere in that collective volume does any author venture a definition of the term. Gardner and Bartlett [22] finally proposed a definition that may stand the test of time. They described the electronic nose as:

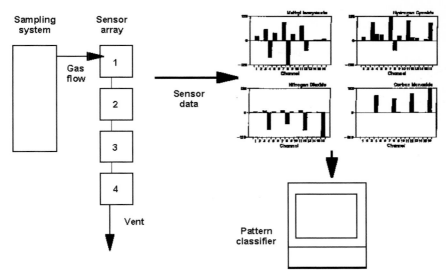

Figure 2.3-3. Fundamental structure of an electronic nose, including a sampling system, an array of sensors, and a pattern classifier. Also shown are typical array responses for different pure chemicals.

'... an instrument, which comprises an array of electronic chemical sensors with partial specificity and an appropriate pattern-recognition system, capable of recognizing simple or complex odors.'

We would interpret the term instrument to include the sampling system and 'partial specificity' as referring to the broad, differential selectivity that is observed in most reactive chemical sensors, such as metal oxide or electrochemical. Some authors, including some of those who are deliberately attempting to model the olfactory system, continue scrupulously to avoid the 'electronic nose' terminology [23, 24]. In fact, it is probably too soon to place a restrictive definition on the electronic nose concept. At this time, a great deal of imagination is being devoted to ways of deriving real-time, multidimensional patterns from vapor samples. The configuration that will prove most general and useful, and which may eventually be called 'the' electronic nose, may very well be still awaiting discovery.

After the sensor array, the second defining component of an electronic nose is the pattern classifier. This is a means of extracting information from the collected sensor array patterns by comparing or associating them. In practice, this means associating a pattern from an unknown sample with a set of patterns from known standards to determine the closest match, for the purpose of identification.

The sampling system is usually sufficiently important that it is listed as the third essential component of an electronic nose. The sampling system assures that samples are supplied to the sensor array in a reproducible way, and conditioned if necessary to adjust concentration, temperature, water vapor concentration, etc. Some sampling systems also provide for automated measurement of a series of samples, although this is not a defining requirement.

2.3.4.2   History of the Electronic Nose

It seems that the electronic nose has had independent origins, in Europe, Asia, and North America. The European effort originated in attempts to mimic the mammalian olfactory system [25]; the North American genealogy had its roots in directed development of field portable identification and quantitation of toxic vapors [26], and the Asian effort to engineering solutions to specific odor detection systems [27]. These separate efforts are detailed in the remainder of this subsection.

To our knowledge, the first attempt to use sensor arrays to emulate and understand mammalian olfaction was carried out by Persaud and Dodd [25], at the University of Manchester Institute of Science and Technology. Their purpose was to model the current conception of the mammalian olfactory system by demonstrating that a few sensors could discriminate among a larger number of odorants. They constructed an array of three metal oxide gas sensors, which they used to discriminate among 20 odorous substances, including essential oils and pure volatile chemicals. Pattern classification was performed by visual comparison of ratios of sensor responses. A decade later, their discovery took the form of the AromaScan®, a commercial 'electronic nose' instrument [28–30].

In 1980, in North America, a group led by the first author started to build an analytical instrument based on a chemical gas sensor array [20, 31–33]. At the time, our goal was to build a portable instrument for the US Coast Guard. This organization needed a portable instrument that would rapidly identify and measure volatile chemical vapors in emergency situations, such as hazardous material spills. Rather than adopting one of the conventional approaches to chemical analysis, such as gas chromatography or infrared spectrometry, we elected to examine the differential responses of an heterogeneous array of chemical sensors. We considered electrochemical, infrared, MOX, catalytic, photoionization, and other candidate sensors. Ultimately, the prototype used only four electrochemical sensors. Since many common chemicals, such as hexane, do not react on electrochemical sensors, a heated catalytic filament in the sample stream converted electrochemically inactive compounds to reactive ones [20]. By varying the temperature of the filament, different patterns were obtained from the four sensors (Figure 2.3-4). This catalytic conversion of the sample was reproducible over wide ranges of concentration and relative humidity, provided the filament temperature and sample flow rate were tightly controlled. Different patterns were obtained from the sensors by operating the filament at a different temperature. By operating the filament at four temperatures from 600 to 950 °C, the number of sensors was effectively increased from 4 to 16. We would say now that the instrument had four physical sensors and 16 virtual sensors. A relatively straightforward application of the 'k-nearest-neighbor' pattern classification algorithm was used to compare the resulting patterns and to identify unknown vapors automatically. We demonstrated that this tiny heterogeneous sensor array could be packaged in a camera bag and accurately identify and quantify over 30 compounds. We estimated that well over 100 compounds was possible [26, 31, 34]. This work resulted in the first engineered, packaged, fully functional field-tested

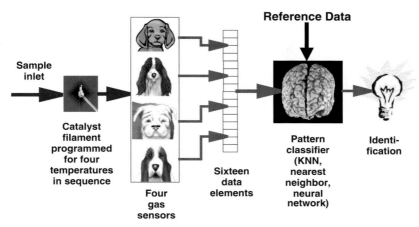

Figure 2.3-4. The 'ILLI-Nose', designed by the authors, which uses four electrochemical sensors and a catalytic converter to generate 16 virtual sensors.

instrument that included both the gas sensor array and pattern classifier. A commercial version was offered for sale in 1985 by Transducer Research Inc. (Naperville, IL). The instrument was recently reconstructed in our laboratory and dubbed the 'ILLI-Nose' (for Illinois). It still rivals modern instruments in selectivity and detection limits.

In the course of this work, we sometimes informally used the term 'electronic nose'. Now that we know more of the operation of Nature's noses, the term is no longer a joke. We can see that there are direct and meaningful parallels between the structure of the natural and artificial noses (Figure 2.3-5). The human nose, for instance, has olfactory sensor cells, each with a receptor protein that interacts with a range of volatile odorant molecules. There are estimated to be 100 million receptor cells, but only 100–200 different types of receptors [35, 36]. The brain learns to recognize the *pattern* of signals associated with certain odors, rather than the response of individual, highly selective cells.

In Asia, scientists were also beginning to foresee the potential of sensor arrays. In 1984, Iwanaga et al. [37] proposed an instrument employing an array of MOS sensors, but suggested using simultaneous equations to compute the relative concentrations of gases in a sample. In order for this approach to work, all the gases present in the sample would have to be known. Clifford [38, 39] patented a similar idea in North America in 1985. This approach does not use pattern recognition and is suitable only for certain industrial processes where all the analyte gases are known. This severely limits the use of the array in uncontrolled situations, ie, most of the proposed uses of the E-nose in food science, medicine, and industry. More complex data interpretation was needed [26].

The work on chemical sensor arrays therefore proceeded on three continents in the 1980s. Ehara et al. [40] suggested an array of microfabricated MOS sensors, but relied on visual comparison of the signals for comparison of samples. In 1986, Ballantine et al. [41] used an array of four SAW sensors to generate pat-

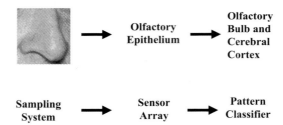

Figure 2.3-5. Functional comparison of the mammalian olfactory system and the electronic nose.

terns, and found structure-activity correlations among a series of substances, including selected chemical warfare agents. Importantly, they employed pattern-recognition methods to treat the array signals. This pattern recognition work was emphasized by Rose et al. [21, 42]. With a sampling system consisting of an organic sorbent preconcentrator and a short GC column, the SAW array was embodied in an instrument that was ultimately offered for sale [43].

About the same time as Persaud and Dodd's work resulted in the commercial AromaScan, we independently addressed the power of sensor arrays to work with ill-defined samples such as odors and flavors. We successfully used our original electrochemical sensor array instrument, with an improved sampling system and neural network pattern recognizer, to classify odors from samples of sound and spoiled grain for the US Department of Agriculture [44]. The final device was nearly as reliable as trained human inspectors at classifying grain samples as 'good', 'sour', or 'musty'. The electronic nose was about 83–96% in agreement with the human inspectors, depending on the sample type. According to data supplied with the samples, the human inspectors only agreed with one another at the same rate of 93%.

The explosion of interest in sensor array technology, especially in Europe, led to the international conference in Iceland [13]. We are humbled by the large number of scientific and engineering groups throughout the world who have made contributions to this development of the electronic nose technology (see reviews cited). All of us in the E-nose field are sincerely indebted to the great volumes of scientific and engineering work in tiny chemical sensors, computer algorithms for pattern classification, low-power portable microcomputer systems, and our understanding of the olfactory process. Without these advances, the E-nose would never have become a reality for any of us.

2.3.4.3   Sensor Array Configurations in Electronic Noses

Like the mammalian olfactory system, the electronic nose uses a holistic approach to distinguish aromas. It does not separate or attempt to identify the individual chemicals responsible for the pattern. The electronic nose works equally well on pure compounds or on undefined samples such as flavors, aromas, and other complex odors. Conventional reductionist analytical methods often become less reliable as sample complexity increases. Our work on grain odors, for example, was preceded by a 3-year effort by the USDA to use GC-mass spectrometry (MS) to distinguish grain quality. The author of the final report on this study stated, after analysis of more than 300 samples, that 'no relationship between the chemical composition and the odor could be found' [45]. Other authors have had similar problems correlating the results of detailed chemical analysis with organoleptic responses [46, 47].

Although conventional analytical instruments become less effective as sample complexity increases, the electronic nose retains its ability to discriminate closely related samples. It has found applications in the food and cosmetics industry, and there are other potential applications everywhere in industry. A particularly potent example is coffee, where the simplest electronic noses can make fine distinctions between blends [48]. Vintners have been able to identify wines by provenance as well as vintage [49, 50]. This distinguishing power can be thought of as an outcome of the chemical imaging character of the array that has been estimated to be able to represent more than 10^{21} different dimensions with even a simple array [51]. This view of the sensor system and its dimensions is a unique consequence of the complexity and selectivity of chemical and biochemical reactions.

Because the electronic nose is not yet a fixed configuration like a gas chromatograph or a mass spectrometer, much imagination has gone into its design. In this section, we will review the following topics:

- the standard or 'homogeneous' sensor array;
- virtual sensor arrays;
- heterogeneous (mixed-class) sensor arrays.

2.3.4.3.1 The Homogeneous Sensor Array

The majority of electronic noses, commercial and otherwise, use sensors of the same class, but generally of different types within a class. Figaro Engineering Inc. lists more than 25 tin oxide sensors in its product line, making a large variety of arrays possible for just this one class. A recent example is the array of 12 tin oxide sensors employed by Romain et al. [52] to discriminate environmental odors. Sensors may be operated under different conditions to alter their selectivities. Tin oxide sensors, for example, can be heated to different temperatures. There are fewer basic types of electrochemical sensors than of tin oxide sensors, but their selectivity can be modified further by altering the working electrode potential.

2.3.4.3.2 Virtual Sensor Arrays

A virtual sensor array consists of a small number of sensors, perhaps only one, which is manipulated to contribute many quasi-independent elements to the output pattern. In our original sensor array instrument, for example, four amperometric sensors were made to yield 16 signals by pretreating the sample over a hot filament at four temperatures [20]. The response of alcohols, which have a strong natural response on amperometric sensors, decreases as the filament temperature is increased. Presumably, the alcohol is oxidized to less reactive substances. Most other organic compounds, such as ketones and alkanes, show increasing responses as temperature increases. Mielle and Marquis [53] recently used a tin oxide sensor operated at seven different temperatures on the same sample to generate quasi-independent pattern elements. Sensitivity to aldehydes was highest at low or intermediate temperatures, and sensitivity to alcohols was highest at the higher temperatures. Strathmann et al. [54] used sample adsorption followed by programmed thermal desorption to preconcentrate sample vapors for the MOSES E-nose. They discovered that they were also able to derive extra pattern elements from desorption at stepped temperatures, greatly increasing the selectivity as well as the sensitivity of the MOSES instrument.

Since pattern classifiers are blind to the physical source of their data, some electronic nose designers have devised even more radical ways of extracting independent pattern elements. Agilent Technologies (formerly Hewlett-Packard) has developed an electronic nose based on a mass spectrometer. Individual m/z peaks, or total ion current integrated over selected mass ranges, have been used to generate the patterns [55]. Another designer uses a single SAW sensor as the detector for a GC column. The heights of selected peaks serve as the pattern data [56]. Company literature advertises '500 sensors', referring to the elution intervals. This is not really new, since chemists have looked for structure-activity relationships in gas chromatograph data for 30 years. Optical spectra, especially in the analytical region of the infrared, have also been used historically to identify compounds; recent technology has involved pattern classification algorithms [57].

Data derived from MS or GC is still subject to the same 'MOSES principle' (see below) restrictions as data from discrete sensors (see the next section). Data sources, whether discrete sensors or a region of a mass spectrum, that do not contribute to the identification of the sample will still add noise. Accordingly, the selection of data sources must be optimized for every application.

2.3.4.3.3 Heterogeneous (Mixed-class) Sensor Arrays

Developers of electronic noses have experimented with arrays of different sizes. The most direct way to improve the data would seem to be to increase the number of sensors of the same class, eg, using 20 sensors instead of four. However, this approach meets rapidly diminishing returns. Stetter et al. [21] demonstrated strong correlations among data in different data array elements from their virtual electrochemical sensor array, presumably because treatment of the sample vapor with the heated catalyst filament did not produce entirely independent data. Similar interdependence might be expected with any array where all sensors belong to the same class. A statistical approach illustrates that a small array of six sensors is all that is required to differentiate up to 100 different patterns/compounds when found alone or in mixtures of up to four of the compounds that constitute the database [34]. However, an assumption restricting the response of each pattern to five of the six sensors had to be made. This illustrates the importance of 'off' or no response data in such array systems.

In 1998, Göpel and colleagues demonstrated that if sensors from several classes are used to form the array, the discriminating power is greatly increased [58]. Their prototype instrument contained eight MOX sensors, eight QMB sensors, and eight calorimetric sensors, arranged in separate and interchangeable modules. They demonstrated greatly increased discrimination between olive oils of different provenance. They were also able to observe the development of rancidity in olive oil samples over time, which is due to the auto-oxidative formation of short-chain aldehydes. As a result of these experiments, the authors enunciated the so-called 'MOSES principle' (where the acronym refers to '*MO*dular *SE*nsor *S*ystem', after the modular design of the instrument):

- As array sizes increase, each additional type of sensor of the same class contributes less information, because sensors within a class are rarely orthogonal. However, each additional sensor contributes the same amount of noise. Therefore, there is an optimum array size.

- The use of different sensor classes, which respond to different physical or chemical properties of the analytes, allows larger numbers of sensors to be used, while still contributing information to the data set.

This approach is not so obscure that it has not occurred to others independently. The sensor array patent of Stetter et al. [33] included a claim covering sensors of mixed classes. Commercial electronic noses made by Alpha MOS (Toulouse), EEV Ltd. (Chelmsford, UK), Nordic Sensor Technologies AB (Linköping, Sweden), RST Rostock Raumfahrt und Umweltschutz GmbH (Rostock, Germany), and Lennartz Electronic GmbH (Tübingen, Germany) later used sensors of mixed type [59].

2.3.4.4   Sampling Systems for Electronic Noses

A sampling system should deliver a vapor sample to a sensor array in a reproducible way. Its purpose is to reduce sample-to-sample variation that may result from differences in humidity, temperature, concentration, etc., as well as to preprocess the sample in any way that increases the quality of output data. To date, little research has been carried out on optimized sampling systems, even though some authors have demonstrated dramatic increases in performance of sensor arrays [60].

The most obvious use of a sampling system is to concentrate the vapors, in order to improve the sensitivity of the sensors. This approach was pioneered by Grate et al. [43] in a four-SAW sensor device which was optimized for chemical warfare agents. It was later commercialized for hydrocarbon measurement by H. Wohltjen at Microsensor Systems Inc. A short tube of organic sorbent (Tenax) was used to absorb vapors from the air. These were desorbed using a heater and passed through a short GC column to the sensor array. Similar preconcentrator approaches have been used by other authors [54].

Samples often contain substances that are common to all and, although the sensors are dominated by them, they do not contribute to discrimination. In bacterial cultures, for example, the common substance is water. In beer and wine, water and alcohol will be present in all samples in much larger quantity than any other constituent. We have successfully used Nafion tubing and anhydrous sodium sulfate to remove selectively water, alcohol, and some other hydrophilic vapors from samples [61]. Although sensor signals, on average, are reduced by a factor of 10 or more, the removal of the dominant constituents greatly improves selectivity. Another type of sampling system [60] allowed TNT adsorbed to silica sand to be detected and discriminated from structurally similar compounds by vaporization of a sample from a tiny beaker, using a hot platinum filament. A second filament was located downstream to combust the sample to electrochemically reactive compounds, probably nitrogen oxides and carbon monoxide.

2.3.4.5   Signal Processing and Pattern Classifiers
for Electronic Noses

Research into pattern classifiers has been perhaps the most aggressively pursued aspect of electronic nose research. A large body of previous research on pattern classification and recognition has been drafted into the quest to extract the maximum information from the chemical data produced by the sensors. We cannot discount the glamor of artificial intelligence methodology, which attracts research effort at the expense of mundane hardware development. Even US funding agencies will readily fund the development of insubstantial mathematical tools rather than support the development of new instruments and sensors.

Nevertheless, two of the chief problems hindering E-nose applications are not being actively addressed. One of these problems is *detection limit*, the ability to detect small amounts of an analyte in a typically responsive matrix. The second problem is the detection of analytes in a *variable matrix*, eg, the detection of the characteristic odors of disease on the breath. Human breath can vary in composition with factors that have nothing to do with disease states, such as diet, smoking history, cultural background, and nutritional state [62], and so there must be some scatter among normal individuals which may complicate the task of disease detection. In principle, the notion of isolating a known class (the disease pattern) from a variable background should be solvable to the same degree as isolating a variable class from a constant background. In either case, we expect some loss in sensitivity, but perhaps not to a degree that would interfere with the power of disease detection. The reason it is not yet solved is because the method for pulling a pattern from the matrix in which the response patterns are most probably related is akin to the brain recognizing that THE CAT can be written 'T_E C_T'. We can read it, especially if it is given in context, but a computer has trouble with or without context. We need to understand how to generate the data that are truly independent and understand the ways or algorithms that allow us to pull the meaning from the data to solve this problem. We are not saying it is impossible, but we are saying that no-one has succeeded yet in solving the two problems of sensitivity and/or selectivity in a variable sample matrix.

Many pattern classifiers have been examined for use in electronic noses. Decision-surface methods are directly applicable to arrays of arbitrary data. In our first explorations of sensor array technology, we used the k-nearest-neighbor (KNN) method [63], which is intuitively understood by people with a minimal mathematical background. It proved to be surprisingly powerful. Years later, we compared its performance directly to a back-propagation artificial neural network (ANN) [44]. Both methods were equivalent in their ability to classify unknown grain samples. However, when either random or systematic error was added to real data, the ANN proved to be much more robust. The rate of successful classification by KNN decreased twice as quickly with increasing error as the ANN method. Moreover, we determined empirically that periodic full calibration was not needed as often with ANN. Partial calibration by known samples related to the unknowns was found to be sufficient.

By far the most popular method of pattern classification in today's electronic noses is principal components analysis (PCA). PCA is primarily a method of reducing the dimensionality of data. An n-sensor array produces n-dimensional data, which cannot be plotted on simple graph paper if $n > 3$. By extracting principal components, most of the systematic variance is typically gathered into two or three dimensions, readily plotted and giving an appearance of comprehensibility. On the other hand, reduction in dimensionality must necessarily result in some loss of information. It nevertheless remains popular, if only because it has the distinctive feature of producing attractive plots that clearly display grouping of replicate samples, and the relative affinity of different samples to one another. These plots have strong presentation, not to mention marketing, appeal; alternative, and more powerful, pattern classifiers do not produce such visually appealing output.

If any lesson has been learned from the use of PCA, it is that one must take care with the questions that one asks of a data set with any statistical treatment. A data set that we gathered from bacterial headspace gases for different periods of growth can be used to understand the most common misuse or error in the application of PCA methods [60]. The data set consists of two bacterial species; separate cultures were grown for 0, 0.5, 1, 2, 3, and 6 h. If PCA is applied to the entire data set in an attempt to separate all the classes at once, the 2 h class can be separated from the 1 h class. That is, growth can be detected at 2 h. If, however, only a portion of the data set is selected, specifically, the 0, 0.5 and 1 h cultures, these time points can readily be discriminated, even by PCA (see Figure

2.3-6). In the first instance, we are implicitly demanding that the statistical method, which draws a mutually perpendicular axis along the dimensions of maximum variance, separate all the classes at once when reduced to the two dimensions. In the second case, we are asking a much less complex question of our data set: can the 0, 0.5, and 1 h samples be separated from one another?

In other words, it is important to be clear about *the question that is being asked of the pattern classification method.* In the first instance, we attempt to separate all the classes at once, first, because it makes a nice illustration for a publication or presentation, and second, because we have not realized that we were even phrasing a question! In the second instance, we have thought about the question that we wanted to ask in the first place: *How early in bacterial growth can we detect differences in the headspace gas?* Accordingly, we pruned the data set to deliberately look at only the relevant data. The rotation of the axes that are used to separate classes in PCA is no longer constrained by the presence of unimportant data. The lesson to be learned from this illustration is that, when you are viewing one of the myriad PCA plots generated for the E-nose data, this representation is only valid and representative for the data set under consideration. No extrapolation to unknowns, and no extrapolation to subsets, can be made. Each extrapolation must be validated, and the accuracy and precision of the extrapolation measured by a valid statistical technique.

Although many arcane advances have been made in the improvement of pattern classification, this subject is outside the scope of this review. We will limit our comments to two promising approaches.

Llobet et al. [64] proposed a pattern classification approach called 'fuzzy ARTMAP', which is able to update its training while not losing its previous training. Using samples of increasing complexity (pure alcohol, coffee, and breath from ketotic cows), the fuzzy ARTMAP method substantially outperformed a conventional back-propagation neural network. The cow's breath samples were taken in a barn, obviously under widely variable environmental conditions. This study of the fuzzy ARTMAP approach is one of the few that has identified and directly addressed the variable matrix problem.

An additional nagging problem that is being attacked by signal processing development has been the problem of calibration. All chemical sensors exhibit drift over time. Much of this is systematic and so can be predicted, at least in part. Although drift is basically a hardware problem, there is no likelihood of solving this problem in the short term, and so some workers have turned to mathematical treatments to deal with it. Holmberg et al. [65] have developed methods to predict future calibration from past drift behavior. Of course, the continuing effort to develop sensors that do not drift or that have drift compensation [66] continues to be welcome since these decrease the reliance on algorithms which cannot entirely compensate for bad data.

2.3.5   Debunking Myths about the E-Nose

This section is presented to stimulate scientific and technical discussions. There are some clarifications required because words and terms can sometimes have ambiguous meanings and sometimes because there is genuine scientific controversy. But truth is found through open dialog and often disagreement is necessary to inspire progress. We sincerely hope that this section leads to better understanding of the E-nose and more rapid progress in its development and application. We apologize to those who do not hold these myths to be self-evident, and we realize that these are in some manner the humble constructs of the authors in order to structure discussion. However, many of these myths are in fact dearly held by potential users of the technology and may serve to warn the uninitiated who are just entering this field of research.

2.3.5.1   Myth 1: The E-Nose is a 'Better' Solution
to Analytical Problems

The analytical capability of the 'electronic nose' has been greatly oversold, and as a result the technology has suffered damage to its credibility in certain applications that has taken years to overcome. Some entrepreneurs and researchers have still not learned the lesson, and continue to tout the E-nose as the final and only solution to virtually every analysis problem. Of the four major classes of analytical problem (Table 2.3-2), the E-nose is only suited for selected subsets of these applications. The analytical problem, at least with today's level of E-nose sophistication, may often be better performed with another analytical method like GC, IR, UV, electrochemical sensor, or GC-MS. The user still deserves the best available solution to their problem and often it will not be the E-nose.

The fact is, the E-nose is not better than conventional techniques like GC or GC-MS, but rather it is *different*. An example would be the detection of the rancidity in olive oil, for which we know the molecular cause, ie, the accumulation of C_5 and C_6 aldehydes over time with exposure to air. After a modestly difficult workup, these aldehydes can be measured directly using GC or GC-MS. Rancid olive can also be discriminated from fresh by the E-nose using the headspace above a stored sample. The sample workup is nonexistent, but you do not get information on precisely what you are measuring. You are only aware that the oil has a pattern that is associated with patterns of rancid oils, and is different from those of fresh oils. For many applications, this is sufficient. Moreover, it is a simpler and less expensive method than GC-MS.

Interestingly, as samples get more complex, and endpoints become less definable in a molecular sense, conventional methods lose their power, but the E-nose retains its ability to discriminate differences between samples. In a 1 year study [44], discrimination of grain samples according to the USDA categories of 'good', 'sour', 'musty', and 'COFO' (Commercially Objectionable Foreign Odor, a 'catchall') with fidelity approaching that of an expert human panel was achieved with an E-nose. A prior study using GC-MS, lasting at least 3 years, had failed to find correlates between organoleptic scores and specific compounds observed in grain samples by GC-MS [45].

Coffee provides an example of an E-nose application that is appropriate to the method. The flavor and odor of coffee are composed of upwards of 640 volatile compounds [13]. Relatively minor changes in components (not necessarily major components, either) result in detectable changes in flavor and odor, according to no discernible pattern. Coffee, on the other hand, is a particularly good subject for E-nose analysis, and is capable of making fine distinctions among coffee varieties and blends [47].

During the 1960s and 1970s, a substantial effort had been spent using GC, GC-MS, and pyrolysis-GC to discriminate among bacteria. It was largely successful, as far as it went, but the complexity of the methods and the continuing need for expert involvement made the methods unfeasible except where no alternatives existed (as in food chemistry), and the methodology has been largely abandoned. The E-nose provides the potential to do bacterial analysis in foods and infectious disease diagnosis with a simplified 'black box' approach which is more amenable to automation, portability, minimal training, and very low cost. A worthwhile goal indeed is our long-term dream to produce a portable breath tester for tuberculosis, which is cheap enough, rugged enough, and simple enough to be used throughout the world, with less reliance on the cold chain or central laboratories [67].

2.3.5.2   Myth 2: The E-Nose Senses or Determines 'Odor' and Works Like the Human Nose

E-nose devices respond to the chemicals to which the sensors respond. The concept of odor is a human one and so human receptors/sensors respond sometimes to totally different compounds than the E-nose sensors. The human sensor responses together with the complex human brain form the substance of a true 'odor.' By its nature, the E-nose most probably is not responding to the same group of compounds as the nose and, in fact, different human noses most assuredly produce different response patterns and are not the same. In this case, the tremendous flexibility of the brain still categorizes/learns the different patterns. However, the E-nose response pattern for the odor and its matrix may also be distinct from all other patterns of the odors and their matrices under consideration. In this case, the E-nose, with its pattern classifier, can 'tell' one 'odor' from another in the data set of odors. In this way, it can be called an odor classifier but should not be confused with identifying a 'human odor' except in enlightened circumstances.

Does the E-nose work like the human nose? In some ways this is of course true and we can see this from the introduction, definition, the above discussion of the EN. But the sensors differ in critical ways. The nasal receptors typically amplify the sensor signals chemically in a manner much unlike current electronically amplified chemical sensors. For example, a single analyte molecule might unblock a channel in a membrane and release 10^5 or more ions. This makes the relative intensities in the nasal receptor patterns extremely large. We need to develop sensors that can chemically amplify (as well as electronically amplify) sensor signals. Perhaps neural nets and fuzzy logic methods of pattern recognition, which begin to simulate reinforcement of differences, are more effective in E-noses for this very reason [44, 64]. While we can learn from nature, we must be prepared to improve upon it, just like the microscope improves upon our eyes when viewing small images and the telescope improves our grasp of far away images. As the eye is not a spectrometer and can be fooled, so too can the nose be fooled. Our new E-noses must be more robust analytical instruments if they are to be deployed widespread.

2.3.5.3 Myth 3: More Sensors in the Array are 'Better'

Are more sensors better? The debate continues, although most parties are now in agreement that there is an optimum number of sensors for each application, although that number and the choice of sensors may differ greatly from one application to another. Some smaller arrays appear to have as much or more capability than large arrays and the theoretical limit of selectivity has not been reached. At least one manufacturer of electronic noses has recognized this. Cyrano Sciences Inc. sells its Cyranose 320 with 32 polymer composite sensors. Specially developed software that isolates those sensors that contribute to the discrimination of samples for each application. These are typically just eight sensors, although the selected sensors differ by application.

Several observations may be made. There is a clear benefit to more sensors if they are completely and perfectly redundant. The theoretical sensitivity of an optical array of completely redundant sensors has been shown to increase as the square root of the number of sensors as expected from simple Gaussian error theory [17]. A second observation is that too many sensors that are not exactly the same but are also not independently responsive or orthogonal can add noise but no new signal information and actually reduce the differentiating power of an array [21, 58]. A third observation is that the potential information content of an array of signals is quite large even for a small array [51, 68, 69]. A fourth observation involves the increased information content of heterogeneous arrays, ie, arrays made with different classes of sensor, as opposed to arrays made from sensors of a single class [58]. This is a direct result of the increased orthogonality of the data from different classes of sensors. At this stage of development of E-nose technology, the selection of the optimum sensor array is still an empirical science.

2.3.5.4   Myth 4: All the Sensors in the Array Need to be 'Partially Sensitive'

Many arrays use sensors that respond to virtually every substance and these seem to work in the E-nose and so, therefore, it might be true that this is needed. For sure, the information content is directly related to the different sensitivities to the same compounds or differential selectivity of the sensors in the array. There is a general assumption that the human nose works because of the relative signals from the receptors. We can agree with all of this. This leads some to conclude that an array of partially sensitive chemical sensors is the only way to make an E-nose. We disagree with this statement without clarification. Just like the eye sees different wavelengths by relative responses of rods and cones, the nose smells by differential responses. But this is not the entire story because the eye can easily be fooled into believing that blue + yellow is the same as green. However, a spectrometer cannot be fooled in this way because it immediately knows if the light is green or a mix of blue and yellow wavelengths. Why? Because the spectrometer is measuring a more fundamental property, the wavelength, and not a group of relative reactions on partially sensitive sensors. This analogy must be true of the nose also. So, admitting that there is a lot to be learned from the nature of the human nose, do we really want a human nose or do we want an instrument that is less ambiguous? We think the goal is the latter. So what are the differences we need and what must we do to get there?

The statement has been made that the E-nose requires a group of 'partially-selective' sensors and they should '*all*' be partially on for each compound in the gas stream. While this may be sufficient for some applications, this is not at all a general requirement but a myth. In fact, arrays that have some 'off' channels should theoretically be better. And, in fact, the human nose probably gets as much information from the receptors that are not on as from the receptor that are on. Calculations illustrate that, using a statistical approach, the minimum number of sensors required to sort any one of 100 compounds into its components of up to a four component admixture was only six gas sensors. In this data set, each analyte had a response on only five of the six sensors [34] and one had to assume one channel was 'off' to get the statistical scheme to work.

A second reason why the array of partially sensitive chemical sensors is incomplete is because we have seen that digital data can improve pattern classification [34]. In this case, intelligence about the signal was used to reduce the number of possibilities in the library, ie, choose from among a smaller set of possibilities [70]. Consider the receptors in the human nose that amplify the signal up to 10^5 times. This has the effect of making the data nearly digital. This receptor response has the effect of emphasizing certain receptors well beyond the noise and it would be as if we knew beforehand certain important sensors and in our software we could weight them heavily before interpretation. It seems we can do this with chemical sensors if principles are chosen that are radically different and heterogeneous. Responses that are 'extremely' sensitive to one or a few compounds greatly improve discrimination power for arrays. Such performance may only come from heterogeneous arrays, eg, an electrochemical sensor will respond to ppb levels of NO, but a QMB will not (it is 'off'), whereas the electrochemical sensor is not responsive to benzene and the QMB can see ppm levels of benzene. Again the empirical observation that heterogeneous arrays fare better in applications supports this discussion.

Our third observation on this point follows from the above. The human nose has receptors that are partially responding but it seems to me that some of them are totally nonresponsive to a given stimulus. The power of the mammalian nose, especially that of the dog, is legendary. Thus, we could predict that the most powerful arrays would be smaller, heterogeneous and contain some on/off channels, channels with orders of magnitude differences in relative sensitivities.

2.3.5.5   Myth 5: It is Easy to Calibrate an E-Nose and Extrapolate to Unknowns

No-one may seriously believe this, but a more pertinent question may be: can calibration can be extrapolated to situations involving unknowns? This is true for any well-developed and validated analytical method, but not many of these exist today for the E-nose technology. Analytical method validation is a specific and thorough process [71]. This step must proceed with caution and is the most difficult part of the application of the E-nose to any real-world problem. Simply put, the issue is that multidimensional sensory data have interferences, drift, and noise that are also multidimensional and adequate methods to handle this situation are not yet easily validated for many applications!

The rules for pattern classification are simply: (1) the pattern of responses must be statistically related to the endpoint, (2) the answer must be able to be adequately represented by the set of responses, (3) a relationship can be discovered by applying the chosen algorithm to the data, and (4) the relationship can be validly extrapolated to additional situations and unknowns. It is up to the E-nose method developer to prove statistically that his endpoint (flavor or odor) is statistically related to the pattern achieved by the sensors. Further, statistical methods strictly apply only to the data set under consideration unless you can also 'prove' a relationship to unknown data as we have discussed above for the PCA plots (Figure 2.3-6).

Having burst the bubble here for some potential users, it is justified to be skeptical of E-nose results. However, can statistically valid relationships been found for difficult analyte/matrix data sets? The answer is a resounding yes. The power of the sensor array to represent feature space is apparently immense and estimate at more than 10^{21} features [34]. But the analyst must perform sufficient tests on calibration sets and unknown sets to understand that they are related and can be represented using the pattern classifier of choice.

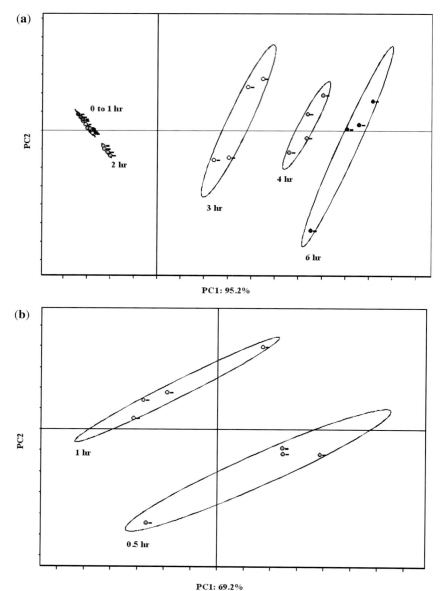

PC1: 69.2%

Figure 2.3-6. (a) Analysis of headspace gases of *Enterobacter aerogenes* cultures. All data are included. Growth at 2 h can be discriminated from 1 h and earlier data, but no growth can be seen at 1 h. Data from eight metal oxide and four electrochemical sensors are included. **(b)** Analysis of the same data set as in **(a)**, except that only data from 0.5 and 1 h are included.

2.3.6   Electronic Nose Applications

As recently as 1998 it was noted [9] that the majority of publications on electronic noses were in the area of research, with only a few 'real-world' applications beginning to appear. Our survey of the E-nose literature, done in February 2000 and again in July 2001, showed a massive increase in the development of applications for the technology. In particular, applications in food technology dominate these.

Sensor array and electronic nose methods have immediate appeal to technologists who must cope with subjective responses of clients and customers to undefined samples. Foods, beverages, cosmetics, packaging materials, consumer polymers, and even wastewaters all have organoleptic restrictions and requirements. Two approaches have been conventionally used to characterize such samples. Subjective analysis by panels of trained persons is the most common approach. Brute-force analysis, in an attempt to isolate the compound(s) responsible for taste and odor, is the other. The Water Department in the City of Chicago, for example, specifically searches for 2-methylborneol and geosmin in drinking water at specific times of year, because these are known to be the cause of the 'earthy' or musty taste of water during algal blooms in Lake Michigan. Electronic noses provide a third, and complementary, approach to characterization. Like taste and odor panels, they are especially useful for comparing samples, rather than relating them to a cause understood at the molecular level.

2.3.6.1   Food

The food industry has spent considerable effort over the years to reduce flavors and odors from an art to a science, with limited success. Many foodstuffs produce volatiles in the concentration range 1–1000 ppm which is appropriate to typical chemical sensors. Not surprisingly, therefore, applications to food have been reviewed regularly since the inception of the electronic nose concept. In 1992, Dodd et al. [72] reviewed some of the potential applications of electronic noses, foreshadowing a large amount of work over the next decade. The subject has since been reviewed by Göpel [73], Kress-Rogers [14], Schaller et al. [46], Krings and Berger [74], Giese [75], and Stephan et al. [76].

Electronic noses have been used for meat, grains, coffee, beer, mushrooms, cheese, sugar, fish, fruits, juices, alcoholic beverages, and packaging materials. Frequently, these reports refer to home-made or early commercial electronic noses such as the AromaScan. Homogeneous arrays with reduced resolving power are usually used. Yet these devices are often capable of the discrimination demanded by the application. Table 2.3-6 lists several references according to the foodstuff examined. Many of these applications are concerned with the detection or confirmation of bacterial spoilage, especially in meat and stored grain. Such studies have sometimes been paired with the study of microorganisms recovered from spoiled meat [77, 78].

Table 2.3-6. References to electronic nose classification of foodstuffs

Class of food		Reference
Fruit	Apples	82, 105, 106
	Pears	107
	Juices	53
	Juices, processing history	108
	Identifying unknown	109
Milk	Off-flavors	104
	Spoilage	89, 92
Cooking oils	Rancidity	81
	Oxidation	110, 111
Grain	Quality	44, 112
	Cereals	113
Meat	General	114
	Beef	78, 115
	Fish	116
	Chicken	79, 80
Beverages	Coffee	13, 101
	Wine	49
	Beer	117
Prepared foods	Mayonnaise	54

A recurring theme in much of the research on food is the correlation of subjective sensory responses, electronic nose responses, and conventional reductionist analysis by GC or GC-MS. Arnold and Senter [79] were able to detect 10 alcohols, from C_2 to C_{14}, an aldehyde, and indole from the headspace of bacteria cultured from processed poultry. In another study, Siegmund and Pfannhauser [80] studied cooked poultry as it became rancid during chilled storage, presumably through nonbacterial spoilage.

Foods, because of their great variety, give us insight into the power of the electronic nose. Some odors are chemically simple. The rancid flavor of old olive oil is due to the accumulation of normal aldehydes [81]. Apples represent a more complex sample. Nakamura et al. [82] cite the flavor of apples as being due to a mixture of nine organic compounds. Commercial artificial apple flavoring is a mixture of these nine compounds. By comparing E-nose results on these mixtures, an optimal sensor array for characterizing apple flavor was selected. Apples also show differences according to variety, integrity, and ripeness [53].

Coffee represents the other end of the spectrum of complexity. The subject of intense experimentation due to its value and complex chemistry, the flavor and odor of this beverage are due to at least 640 volatile compounds [72]. Over 120 of these compounds can be unambiguously identified at one time in a single analysis after extraction by solid-phase microextraction and GC [83]. Much more protracted analysis is needed to determine most of the compounds that actually contribute to flavor and odor. Relatively minor changes in this complex mixture can be readily detected by the senses.

2.3.6.2 Bacteriological

Bacteriologists have long known that many species of bacteria can be identified by their odors [84]. Historically, much of practical bacterial taxonomy has depended on the unique metabolic characteristics of each species of microorganisms. No organism completely oxidizes all its food to carbon dioxide, and it is safe to say that every species of animal has a metabolically determined baseline emission of waste. Since many metabolic products are volatile, it is not surprising that bacterial cultures are good subjects for electronic nose evaluation [44]. Early research in headspace analysis of cultures and medical specimens has been used to identify the microorganisms involved [85–87]. Several authors have demonstrated that bacterial species can almost always be discriminated by even simple electronic noses [88–92].

For practical medical applications, however, detection limit has always been an important issue. McEntegart et al. [61] demonstrated that *E. coli* in a culture medium was not detected until cell concentrations had exceeded 10^8/mL.

2.3.6.3 Medical

The medical field offers great potential for the application of electronic noses. In essence, the human body processes nonvolatile, mainly macromolecular foods into volatile products. Both the normal functioning of the body, and aberrations of that functioning, should be detectable on E-noses. There is a ready market for diagnostic methods that are noninvasive and inexpensive; the E-nose offers both advantages. The medical potential has been addressed recently in an excellent review by Pavlou and Turner [93].

Breath tests, in particular, hold special promise, because of the ease with which samples can be taken. Some research has indicated that the organic volatiles found in human breath do not vary greatly among individuals. Kratoszynski et al. [62] measured over 100 volatile compounds in the breath of healthy people selected across bounds of gender, race, and socioeconomic status. Three compounds, acetone, isoprene, and acetonitrile, made up 50% of the total mass of exhaled organics. Each organic compound varied within an order of magnitude among individuals. These results imply that the variation in matrix between individuals will not be very great. Similar studies have been done by Jansson and Larsson [94] and Phillips [95].

Some of the suggested applications are:

- detection of urinary tract infections from urine headspace gas;
- breath test for cancer;
- breath tests for respiratory infections (staphylococcus, bacterial pneumonia, tuberculosis, etc.);
- breath tests for various forms of poisoning;
- breath and urine tests for metabolic disorders, including solid and humoral tumors and diabetes;
- melanomas, which have been detected by trained dogs using odor (news item);
- wound infections;
- neonatal complications, such as jaundice, where even such simple diagnostic methods as blood sampling are invasive.

Di Natale and D'Amico [96] have recently reported discrimination among patients with lung cancer, nondiseased controls, and lung cancer patients after surgery. Solid tumors are often poorly vascularized and have a metabolism adapted to rapid growth rather than static function. The combination of necrotic or anaerobic tissue and rapid metabolism would be expected to generate volatiles. The Greeks knew cancer as 'the stinking disease', which suggests that other types of cancer will also respond to diagnosis and monitoring by E-nose.

Because of the ease with which bacteria can be detected and discriminated from one another, one would also expect that infections would be easily detected against the background of normal metabolism. Besides volatiles produced by the bacteria themselves, local tissue inflammation and damage caused by the infection would also produce volatile substances. One researcher has reported the detection of bacterial pneumonia, albeit in intubated (ie, seriously ill) patients [97]. Urinary tract infections have been observed to cause compositional changes in the headspace gas above urine specimens [86, 98, 99].

2.3.6.4   Environment and Safety

Analytical chemistry in safety and environment applications usually involves the detection and measurement of pure chemicals such as toxic gases or solvents. Although electronic noses have been used to measure or identify pure chemicals, these instruments are rarely the best approach, compared with such conventional techniques as gas chromatography or photoionization. There are distinct areas, however, where the special properties of the E-nose are helpful.

Rose-Pehrsson et al. [100] have been developing a fire detector for naval and space use that fuses the outputs of many sensors, both physical and chemical. In space, combustion has special characteristics due to the absence of gravity. Flames are often invisible, and carbon monoxide concentrations can increase very rapidly, since fumes do not escape from the fire by convection. Hence it is in the interest of the designer to use as many modes of detection as possible, including ultraviolet and acoustic sensors, as well as carbon monoxide and other gas sensors, in a fire detection system.

E-noses have also been used in the characterization of wastewater. Rather than detailed analysis, wastewater is often characterized by comprehensive methods with endpoints that are as difficult to justify as the E-nose array. BOD (biological oxygen demand) measures the consumption of oxygen by organisms in a sample of water over a fixed time period. Since the composition of such waters can be complex, in certain cases a sensor array pattern can substitute for analysis for several different parameters. Singh et al. [101] used an E-nose to monitor water quality. Later, Fenner and Steutz [102] used the technique to monitor waste treatment.

2.3.7 The Future of Sensor Arrays and Electronic Noses

It is always dangerous to predict the future, but it is often the way we challenge ourselves with the most difficult goals. The development of the electronic nose is still in its early stages. Many instruments have been rushed to market with insufficiently developed sampling systems and especially with sensors that are inappropriate for the task. Elaborate sampling and data-processing systems are typically used to attempt to compensate for the effects of temperature, humidity, memory effects, and low sensitivity. But even the most powerful pattern classifiers cannot operate with low-quality sensor data. Any new technology must overcome a barrier of suspicion, but the modest performance of the early commercial electronic noses has raised the barrier still further. One critic has already, if prematurely, dismissed the electronic nose in print as 'a thing of the past' [103] and another has more accurately reported, '... the long-term performance of electronic noses has not lived up to expectations ...' [104]. Each of the three components of the electronic nose is subject to improvement, to achieve improved sensitivity with minimum impact on selectivity.

Where do we go from here? It is clear that the E-nose has made much progress and come a long way. But there is yet much to do. Our human nose is elegant but not foolproof. It still has more redundancy and self-amplified sensors and is better at most things than an E-nose. This is proven by the existence of the olfactory panels that have not been replaced by the E-nose. However, the sensor array does not fatigue as easily, can be placed in hazardous atmospheres, is less costly, and can travel easily into outer space. It also holds the promise of being much cheaper, smaller and easier to use and maintain than a mass spectrometer. Students have built E-noses for us from spare parts found in the laboratory. When this simplicity is combined with the power of the sensor array, the E-nose warrants the attention it is getting.

If the issues of calibration, extrapolation to unknown data sets, and stability of sensors and patterns can not be improved, the E-nose will never achieve its rightful place in the arsenal of analytical tools. On the other hand, progress on these technical problems offers the promise to revolutionize analytical chemistry in the field. It ill boost the applications of chemical sensors and microinstrumentation manyfold.

Where will these improvements lie? Certainly, at least in part in improved chemical sensors. We continue to observe improvements in drift correction [66]. Such on-board automated compensation, now routine for physical sensors, will become more routine for chemical sensors and spur more applications. Further, chemical sensors will more mimic nature in being self-amplifying and regenerating, perhaps by incorporating biological components and mechanisms. Chemical sensors will be tuned to measure such fundamental chemical parameters as solubility and binding constants, making them more like physical sensors that do not simply measure color but rather wavelength of light. Another area of improvement will be in the array itself with compensation for patterns and additional sensor heterogeneity. Arrays may become heterogeneous in sensor class and type as well as in integration into μ-TAS, with the addition of sampling and separation systems. We can envision data from arrays of arrays that are spatially and temporally separated like the human nose to be an improvement. And, of course, pattern classification must be improved. A tiny bird, whose brain contains the smallest amount of gray matter, can recognize a mouse 1000 feet away in milliseconds, against any background (matrix) while flying, yet the most powerful of our PCs struggle with elementary pattern recognition. We require novel approaches to these algorithms and stacked and/or sequential application of specialized algorithms seem appropriate here by the above analogy.

We hope that these discussions and citations do not offend anyone because of their less than comprehensive nature. One always makes choices in these matters and the responsibility lies with the authors. However, we hope this is a contribution to those in and allied to this field of work.

2.3.8   References

[1]   Seiyama, T., Fueki, K., Shiokawa, J., Suzuki, S. (eds.), *Proceedings of International Meeting on Chemical Sensors*, 19–22 September 1983, Fukuoka, Japan; Amsterdam: Elsevier, 1983.

[2]   http://www.electrochem.org and http://www.ieee.org, nose.uia.ac.be/review/default.asp.

[3]   DeCastro, M., Stetter, J. R., Kwong, L., Penrose, W. R., in: *Proceedings of 8th International Meeting on Chemical Sensors*, Basel, Switzerland, 2–5 July 2000; p. 536.

[4]   Hierlemann, A., Schweizer-Berberich, M., Weimar, U., Kraus, G., Pfau, A., Göpel, W., in: *Sensors Update: Sensor Technology – Applications – Markets*, Baltes, H., Göpel, W., Hesse, J. (eds.); Weinheim: VCH, 1996, Vol. 2, pp. 119–180.

[5] Pearce, T.C., Gardner, J.W., Göpel, W., in: *Sensors Update*, Baltes, H., Göpel, W., Hesse, J. (eds.), Weinheim: VCH, 1996, Vol. 3, pp. 61–130.

[6] Vaihinger, S., Göpel, W., in: *Sensors: A Comprehensive Survey*, Göpel, W., Hesse, J., Zemel, J.N. (eds.); Weinheim: VCH, 1991, Vol. 2, pp. 191–237.

[7] Albert, K.J., Lewis, N.S., Schauer, C.L., Sotzing, G.A., Stitzel, S.E., Vaid, T.P., Walt, D.R., *Chem. Rev.* **100** (2000) 2595–2626.

[8] Jurs, P.C., Bakken, G.A., McClelland, H.E., *Chem. Rev.* **100** (2000) 2649–2678.

[9] Dickinson, T.A., White, J., Kauer, J.S., Walt, D.R., *Trends Biotechnol.* **16** (1998) 250–258.

[10] Nagle, H.T., Gutierrez-Osuna, R., Schiffman, S.S., *IEEE Spectrum* Sept. (1998) 22–38.

[11] Gardner, J.W., Bartlett, P.N., *Electronic Noses: Principles and Applications;* Oxford: Oxford University Press, 1999.

[12] http://nose.uia.ac.be/review/default.asp.

[13] Gardner, J.W., Bartlett, P.N., in: *Sensors and Sensory Systems for an Electronic Nose: Proceedings of NATO Symposium, Reykjavik, Iceland, 1991*, Gardner, J.W., Bartlett, P.N. (eds.); Dordrecht: Kluwer, 1992.

[14] Kress-Rogers, E. (ed.), *Handbook of Biosensors and Electronic Noses: Medicine, Food, and the Environment;* Boca Raton, FL: CRC Press, 1997.

[15] Stetter, J.R., Penrose, W.R. (eds.), *Artificial Chemical Sensing: Proceedings of the Eighth International Symposium on Olfaction and the Electronic Nose (ISOEN 2001), March 26–28, 2001, Washington DC*; Pennington, NJ: Electrochemical Society, 2001.

[16] Gardner, J.W., Bartlett, P.N., *Electronic Noses: Proceedings of International Symposium on Olfaction and the Electronic Nose*; Oxford: Oxford University Press, 1999.

[17] Dickinson, T.A., Michael, K.L., Kauer, J.S., Walt, D.R., *Anal. Chem.* **71** (1999) 2192–2198.

[18] Rakow, N.A., Suslick, K.S., *Nature* **406** (2000) 710–713.

[19] *Proceedings of Transducers '01, 11th International Conference on Solid-State Sensors and Actuators*, Munich, Germany, 10–14 June 2001, pp. 1672 ff.

[20] Stetter, J.R., Zaromb, S., Findlay, M.W., Jr., *Sens. Actuators* **6** (1984) 269–288.

[21] Stetter, J.R., Jurs, P.C., Rose, S.L., *Anal. Chem.* **58** (1986) 860–866.

[22] Gardner, J.W., Bartlett, P.N., *Sens. Actuators B* **18/19** (1994) 211–220.

[23] Persaud, K.C., Travers, P.J., in: *Handbook of Biosensors and Electronic Noses: Medicine, Food, and the Environment*, Kress-Rogers, E. (ed.); Boca Raton, FL: CRC Press, 1997, pp. 563–592.

[24] Byun, H.-G., Persaud, K.C., Kim, J.-D., Lee, D.-D., in: *Proceedings of 6th Annual Symposium on Olfaction and the Electronic Nose*, Tübingen, Germany, 20–22 September 1999; p. 237.

[25] Persaud, K., Dodd, G., *Nature* **299** (1982) 352–355.

[26] Stetter, J.R., Zaromb, S., Penrose, W.R., Findlay, M.W., Jr., Otagawa, T., Sincali, A.J., in: *Proceedings of 1984 Hazardous Material Spills Conference*, 9–12 April 1984, Nashville, TN; pp. 183–194.

[27] Ikegami, A., Kaneyasu, M., in: *TRANSDUCERS'85, Proceedings of the International Conference on Solid-state Sensors and Actuators*; Piscataway, NJ: IEEE, 1985.

[28] Persaud, K.C., Pelosi, P., in: *Sensors and Sensory Systems for an Electronic Nose, Proceedings of NATO Symposium*, Gardner, J.W., Bartlett, P.N. (eds.), Reykjavik, Iceland, 1991; Dordrecht: Kluwer, 1992, pp. 237–256.

[29] Hatfield, J. V., Neaves, P., Hicks, P. J., Persaud, K., Travers, P., *Sens. Actuators B* **18/19** (1994) 221–228.
[30] Amrani, M. E., Dowdeswell, R. M., Payne, P. A., Persaud, K. C., *Sens. Actuators B* **44** (1997) 512–516.
[31] Stetter, J. R., Penrose, W. R., Zaromb, S., Christian, D., Hampton, D. M., Nolan, M., Billings, M. W., Steinke, C., Otagawa, T., Stull, O., in: *Proceedings of Second Annual Technical Seminar on Chemical Spills*, Environment Canada, 5–7 February, 1985, Toronto.
[32] Stetter, J. R., Zaromb, S., Findlay, M. W., *US Pat. 5 055 266*, 1991.
[33] Stetter, J. R., Zaromb, S., Penrose, W. R., *US Pat. 4 670 405*, 1987.
[34] Zaromb, S., Stetter, J. R., *Sens. Actuators* **6** (1984) 225–243.
[35] Mombaerts, P., *Science* **286** (1999) 707–715.
[36] Mori, K., Nagao, H., Yoshihara, Y., *Science* **286** (1999) 711–715.
[37] Iwanaga, S., Sato, N., Isegami, A., Isogai, Noro, T., Arima, H., *US Pat. 4 457 161*, 1984.
[38] Clifford, P. K., *US Pat. 4 542 640*, 1985.
[39] Clifford, P. K., in: *Proceedings of the International Meeting on Chemical Sensors*, Seiyama, T., Fueki, K., Shiokawa, J., Suzuki, S. (eds.); Fukuoka, Japan 19–22 September 1983; Tokyo: Kodansha/Elsevier, pp. 153–158.
[40] Ehara, K., Koizumi, T., Wakabayashi, Y., *US Pat. 4 770 027*, 1988.
[41] Ballantine, D. S., Jr., Rose, S. L., Grate, J. W., Wohltjen, H., *Anal. Chem.* **58** (1986) 3058.
[42] Rose-Pehrsson, S. L., Grate, J. W., Ballantine, D. S., Jr., Jurs, P. C., *Anal. Chem.* **60** (1988) 2801.
[43] Grate, J. W., Rose-Pehrsson, S. L., Venezky, D. L., Klusty, M., Wohltjen, H., *Anal. Chem.* **65** (1993) 1868.
[44] Stetter, J. R., Findlay, M. W., Jr., Schroeder, K. M., Yue, C., Penrose, W. R., *Anal. Chim. Acta* **284** (1993) 1–11.
[45] Weinberg, D. S., *USDA Report SORI-EAS-86-1208*, 1986.
[46] Schaller, E., Bosset, J. O., Escher, F., *Food Sci. Technol.* **31** (1998) 305–316.
[47] Zhou, M., Robards, K., Glennie-Holmes, M., Helliwell, S., *J. Agric. Food Chem.* **47** (1999) 3946–3953.
[48] Gardner, J. W., Shermer, H. V., Tan, T. T., *Sens. Actuators B* **6** (1992) 71–75.
[49] Guadarrama, A., Fernández, J. A., Íñiguez, M., Souto, J., de Saja, J. A., *Anal. Chim. Acta* **411** (2000), 193–200.
[50] Sawyer, A., *Wine Bus. Weekly*, July (1997).
[51] Göpel, W., *Sens. Actuators B* **52** (1998) 125–142.
[52] Romain, A.-C., Nicolas, J., Wiertz, V., Maternova, J., Andre, P., *Sens. Actuators B* **62** (2000) 73–79.
[53] Mielle, P., Marquis, F., *Sens. Actuators B* **76** (2001) 470–476.
[54] Strathmann, S., Hahn, S., Weimar, U., in: *Artificial Chemical Sensing: Proceedings of the Eighth International Symposium on Olfaction and the Electronic Nose (ISO-EN 2001)*, Stetter, J. R., Penrose, W. R. (eds.), 26–28 March 2001, Washington, DC; Pennington, NJ: Electrochemical Society, pp. 48–53.
[55] http://www.chem.agilent.com/Scripts/PDS.asp.
[56] http://www.estcal.com/TechPapers/ASA2000.pdf.
[57] Varmuza, K., *Pattern Recognition in Chemistry*, Springer, Berlin, 1980.
[58] Mitrovics, J., Ulmer, H., Weimar, U., Göpel, W., *Acc. Chem. Res.* **31** (1998) 307–315.

[59] Nagle, H.T., Gutierrez-Osuna, R., Schiffman, S.S., *IEEE Spectrum*, Sept. (1998) 22–38.
[60] Stetter, J.R., Strathmann, S., McEntegart, C., Decastro, M., Penrose, W.R., *Sens. Actuators B* **69** (2000) 410–419.
[61] McEntegart, C.M., Penrose, W.R., Strathmann, S., Stetter, J.R., *Sens. Actuators B* **70** (2000) 170–176.
[62] Kratoszynski, B., Gabriel, G., O'Neill, H., *J. Chromatogr. Sci.* **15** (1977) 239–244.
[63] Tou, J.T., Gonsalez, R.C., *Pattern Recognition Principles*; Reading, MA: Addison-Wesley, 1974.
[64] Llobet, E., Hines, E.L., Gardner, J.W., Bartlett, P.N., Mottram, T.T., *Sens. Actuators B* **61** (1999) 183–190.
[65] Holmberg, M., Davide, F.A.M., Di Natale, C., D'Amico, A., Winquist, F., Lundstrom, I., *Sens. Actuators B* **42** (1997) 185–194.
[66] Cole, M., Gardner, J.W., Bartlett, P.N., in: *Artificial Chemical Sensing: Proceedings of the Eighth International Symposium on Olfaction and the Electronic Nose (ISO-EN 2001)*, Stetter, J.R., Penrose, W.R. (eds.); 26–28 March 2001, Washington, DC; Pennington, NJ: Electrochemical Society, pp. 117–120.
[67] Stetter, J.R., Penrose, W.R., Kubba, S., Kocka, F., McEntegart, C.M., Roberts, R.R., Iademarco, M.F., in: *Artificial Chemical Sensing: Proceedings of the Eighth International Symposium on Olfaction and the Electronic Nose (ISOEN 2001)*, Stetter, J.R., Penrose, W.R. (eds.); 26–28 March, 2001, Washington, DC; Pennington, NJ: Electrochemical Society, pp. 54–61.
[68] Zellers, E.T., Park, J., Hsu, T., Groves, W.A., *Anal. Chem.* **70** (1998) 4191–4201.
[69] Zellers, E.T., Batterman, S.A., Han, M., Patrash, S.J., *Anal. Chem.* **67** (1995) 1092–1106.
[70] Zaromb, S., Battin, R., Penrose, W.R., Stetter, J.R., Stamoudis, V.C., Stull, J.O., in: *Proceedings of the 2nd International Meeting on Chemical Sensors*, Aucouturier, J.L. (ed.); Bordeaux, France, July 7–10, 1986, pp. 739–742.
[71] US Environmental Protection Agency, *EPA Guidance for Quality Assurance Project Plans. EPA/600/R-98/018*, 1998 (http://www.epa.gov/quality/qs-docs/g5-final.pdf).
[72] Dodd, G.H., Bartlett, P.N., Gardner, J.W., in: *Sensors and Sensory Systems for an Electronic Nose. Proceedings of NATO Symposium*, Gardner, J.W., Bartlett, P.N. (eds.); Rejkjavik, Iceland; Dordrecht: Kluwer, 1992, pp. 1–11.
[73] Göpel, W., *presented at the Food and Science Conference*, Heilbronn, Germany, November 1997.
[74] Krings, U., Berger, R.G., *Appl. Microbiol. Biotechnol.* **49** (1998) 1–8.
[75] Giese, J., *Food Technol.* **54** (2000) 96–100.
[76] Stephan, A., Bücking, M., Steinhart, H., *Food Res. Int.* **33** (2000) 199–209.
[77] Rossi, V., Talon, R., Berdague, J.-L., *J. Microbiol. Methods* **24** (1995) 183–190.
[78] Vernat-Rossi, V., Garcia, C., Talon, R., Denoyer, C., Berdague, J.-L., *Sens. Actuators B* **37** (1996) 43–48.
[79] Arnold, J.W., Senter, S.D., *J. Sci. Food Agric.* **78** (1998) 343–348.
[80] Siegmund, B., Pfannhauser, W., *Z. Lebensm. Unters. Forsch. A* **208** (1999) 336–341.
[81] Aparicio, R., Rocha, S.M., Delgadillo, I., Morales, M.T., *J. Agric. Food Chem.* **48** (2000) 853–860.
[82] Nakamura, K., Suzuki, T., Nakamoto, T., Morizumi, T., *IEEE Trans. Electron.* **E83-C** (2000) 1051–1056.
[83] Sanz, C., Ansorena, D., Bello, J., Cid, C., *J. Agric. Food Chem.* **49** (2001) 1364–1369.
[84] Omelianski, V.L., *J. Bacteriol.* **8** (1923) 393–419.

[85]   Hayward, N.J., Jeavons, T.H., Nicholson, A.J.C., Thornton, A.G., *J. Clin. Microbiol.* **6** (1977) 187–194.

[86]   Hayward, N.J., Jeavons, T.H., *J. Clin. Microbiol.* **6** (1977) 202–208.

[87]   Vergnais, L., Masson, F., Montel, M.C., Berdagué, J.L., Talon, R., *J. Agric. Food Chem.* **46** (1998) 228–234.

[88]   Gibson, T.D., Prosser, O., Hulbert, J.N., Marshall, R.W., Corcoran, P., Lowery, P., Ruck-Keene, E.A., Heron, S., *Sens. Actuators B* **44** (1997) 413–422.

[89]   Holmberg, M., Gustafsson, F., Hornsten, E.G., Winquist, F., Nilsson, L.E., Ljung, L., Lundstrom, I., *Biotechnol. Tech.* **12** (1998) 319–324.

[90]   Payne, P.A., Persaud, K.C., *US Pat. 5807701*, 1998.

[91]   Gardner, J.W., Craven, M., Dow, C., Hines, E., *Meas. Sci. Technol.* **9** (1998) 120–127.

[92]   Magan, N., Pavlou, A., Chrysanthakis, I., *Sens. Actuators B* **72** (2001) 28–34.

[93]   Pavlou, A.K., Turner, A.P.F., *Clin. Chem. Lab. Med.* **38** (2000) 99–112.

[94]   Jansson, B.O., Larsson, B.T., *J. Lab. Clin. Med.* **74** (1969) 961–966.

[95]   Phillips, M., *US Pat. 6221026* 2001.

[96]   Di Natale, C., D'Amico, A., in: *Artificial Chemical Sensing: Proceedings of the Eighth International Symposium on Olfaction and the Electronic Nose (ISOEN 2001), Stetter, J.R., Penrose, W.R. (eds.), 26–28 March 2001, Washington, DC; Pennington, NJ: Electrochemical Society, pp. 48–53.*

[97]   Hanson, C.W. III, Steinberger, H.A., *Anesthesiology* **87** (1997) A269.

[98]   Armstrong, D.W., Schneiderheinze, J.M., *Anal. Chem.* **72** (2000) 4474–4476.

[99]   Burke, D.G., Halpern, B., Malegan, D., McCalms, E., Danks, D., Schlesinger, P., Wilken, B., *Clin. Chem.* **29** (1983) 1834–1838.

[100] Rose-Pehrsson, S., Hart, S., Hammond, M., Gottuk, D., Wong, J., Wright, M., in: *Artificial Chemical Sensing: Proceedings of the Eighth International Symposium on Olfaction and the Electronic Nose (ISOEN 2001),* Stetter, J.R., Penrose, W.R. (eds.), 26–28 March 2001, Washington, DC; Pennington, NJ: Electrochemical Society, pp. 206–211.

[101] Singh, S., Hines, E.L., Gardner, J.W., *Sens. Actuators B* **30** (1996) 185–190.

[102] Fenner, R.A., Steutz, R.M., *Water Environ. Res.* **71** (1999) 282–289.

[103] Armstrong, N.W., *Today's Chemist at Work,* October (1999) 9.

[104] Marsili, R.T., *J. Agric. Food Chem.* **47** (1999) 648–654.

[105] Hines, E.L., Llobet, E., Gardner, J.W., *Electronics Letters* **35** (1999) 821–823.

[106] Young, H., Rossiter, K., Wang, M., Miller, M., *J. Agric. Food Chem.* **47** (1999) 5173–5177.

[107] Oshita, S., *Computers and Electronics in Agriculture* **26** (2000) 209–216.

[108] Shaw, P.E., Goodner, K.L., Bazemore, R., Nordby, H.E., Widmer, W.W., *Lebensm.-Wiss. u. Technol.* **33** (2000) 331–334.

[109] Kondoh, J., Shiokawa, S., *Jap. J. Appl. Phys.* **33** (1994) 3095–3099.

[110] Shen, N., Duvick, S., White, P., Pollak, L., *J. Am. Oil Chem. Soc.* **76** (1999) 1425–1429.

[111] Ali, Z., O'Hare, W.T., Theaker, B.J., in: *Artificial Chemical Sensing: Proceedings of the Eighth International Symposium on Olfaction and the Electronic Nose (ISOEN 2001),* Stetter, J.R., Penrose, W.R. (eds.), 26–28 March 2001, Washington, DC; Pennington, NJ: Electrochemical Society, pp. 187–193.

[112] Magan, N., Evans, P., *J. Stored Products Res.* **36** (2000) 319–340.

[113] M. Zhou, K. Robards, M. Glennie-Holmes, S. Helliwell, *J. Agric. Food Chem.* **47** (1999) 3946–3953.

[114] Braggins, T., http://www.nzmeat.co.nz/FILES/dop9798a.htm, 1997

[115]   Blixt, Y., Borch, E., *Int. J. Food Microbiol.* **46** (1999) 123–134.
[116]   Ólafsdóttir, G., Högnadóttir, Á., Martinsdóttir, E., Jónsdóttir, H. *J. Agric. Food Chem.* **48** (2000) 2353–2359.
[117]   Pearce, T.C., Gardner, J.W., Friel, S., Bartlett, P.N., Blair, N., *Analyst* **118** (1993) 371–377.

List of Symbols and Abbreviations

Abbreviation	Explanation
ANN	artificial neural network
BOD	biological oxygen demand
COFU	commercially objectionable foreign odor
GC	gas chromatography
IMCS	International Meeting on Chemical Sensors
IR	infrared
KNN	*k*-nearest neighbor
MIS	metal insulator semiconductor
MOS	metal oxide semiconductor
MOSES	modular sensor system
MOX	metal oxide
MS	mass spectrometry
PCA	principal component analysis
QMB	quartz crystal microbalance
SAW	surface acoustic wave
μ-TAS	micro-total analytical system

2.4 Sensors for Distance Measurement and Their Applications in Automobiles

J. OTTO, University for Applied Sciences, Aalen, Germany

Abstract

Distance measurement under difficult environmental conditions is essential for automotive systems. Vehicles in the near future are intended to drive autonomously or at least to give the driver as much assistance as possible to increase safety and comfort. Distance measurement systems already exist that allow the implementation of systems such as collision warning and autonomous intelligent cruise control. The radar distance measurement used for this purpose works under arbitrary weather conditions, where other measurement principles such as laser, optical, infrared or ultrasonic systems may fail. Microwave radar is the most important measurement method for automotives. Ultrasonic systems serve as a parking aid. Other systems such as video-based systems are under development for autonomous driving within a lane. This chapter describes several distance measurement systems and their implementation in automobiles.

Keywords: Automotive; distance measurement; radar distance measurement; collision warning; parking aid; autonomous intelligent cruise control

Contents

2.4.1   Introduction

Improving the functionality of automotive systems is a major challenge. Starting at an already very high level, the manufacturers of automobiles want to give the driver of a car the possibility of further assistance in addition to increasing the comfort and safety of all passengers, thus improving the market performance of their products. Automotive electronics are of increasing importance in modern vehicles. A large part of these issues are coupled with modern sensors for distance measurement, which are the topic of this chapter. Collision warning, collision avoidance, lane departure warning, parking aids, automatic cruise control (ACC) and autonomous intelligent cruise control (AICC) are new items for modern automotive systems which are possible only on the basis of reliable distance sensors [1, 2].

The whole field of automotive safety development has accelerated in Europe since the 1970s by European collaborative research programmes such as PROMETHEUS, DRIVE, PROMOTE and AC-ASSIST. Many sensors and system concepts have been designed and tested. Automotive companies worldwide are competing with great effort to establish new items of this kind.

One example is the Mercedes-Benz Distronic system being built into their S-class saloon cars since June 1999. This system uses a new 76 GHz microwave sensor. It works not only as a cruise control to fulfil the function of automatic speed control but also checks the distance to vehicles in front of the car. With conventional cruise control a desired speed is selected by the driver, and a control system operates on the throttle to maintain this desired speed without knowing whether there is a slow vehicle or an obstacle in the lane. When traffic density is moderate or high, the driver will have to adjust speed frequently and also will often have to brake, disengaging the cruise control. Thus conventional cruise control can become a source of irritation when used in moderate or heavy traffic.

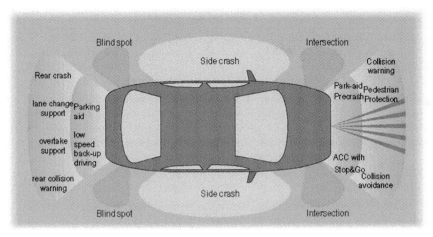

Figure 2.4-1. Different issues for distance sensors in a car (courtesy of Siemens).

Modern cruise controls such as Distronic automatically adapt to the velocity of a vehicle in front and change into the follow mode at a safe distance as long as a car stays in front in the lane. As soon as the lane is free again the AICC switches from the follow mode to the velocity set by the driver. Thus a very comfortable, safe and relaxing way of driving is achieved [3].

Another example is Mazda's second-generation model of its Advanced Safety Vehicle, an experimental car with great efforts to detect and protect pedestrians. The adapted technologies include systems for giving information or warnings to the driver, collision avoidance, and reducing damage from injury. The pedestrian warning system uses a two-dimensional scanning laser radar, which can detect pedestrians from a distance of more than 45 m, even when they are in black clothes at night. Furthermore, the system can distinguish a pedestrian who is standing at the side of the road from one who is starting to cross the road. Additionally an infrared camera, which is mounted under the headlamp, enables the driver to recognize the conditions ahead of the vehicle even in bad weather. Pedestrians, vehicles, or other objects can be detected from a distance of more than 100 m even in fog at night [4].

Other approaches concern automatic lane control and even automatic driving within convoys, which result from increasing traffic density. Here again distance measurement to preceding vehicles or white lines on the road are key issues, employing optical or infrared cameras or scanning devices together with video processing methods. But also within a vehicle distance measurement may open up new comfort and safety aspects, eg, concerning springing, tire control and adaptive airbag function.

Figure 2.4-1 gives an overview of different issues for distance sensors in a car. Surround sensing in all directions has great potential for future applications: in addition to obstacle detection and collision warning at the front and rear of the car, there is the elimination of blind spots and improved airbag function in the case

Table 2.4-1. Distance measurement methods and applications in vehicles

Sensor type	Characteristics/details	Purpose	References
24 GHz radar	Short range (about 7 m), wide angle (at least 110°)	Near obstacle detection, parking aid	[5]
60 and 76 GHz pulse radar	Wide range (up to 200 m), mounted in front bumper	AICC, far-field multitarget detection	[2]
60 and 76 GHz FMCW radar	Low power (<1 mW) Bandwidth: 220–300 MHz Several radar beams	AICC, far-field multitarget detection	[6–8]
60 and 76 GHz FSK radar	Small bandwidth	AICC, far-field multitarget detection	[3]
76 GHz radar imaging	High-resolution pulse radar, images range vs. azimuth	Far distance imaging	[9]
Infrared scanning system/IR laser radar	Optical wavelengths range up to 120 m Small beam divergence (1° vertical, 2° horizontal)	AICC, front and rear obstacle detection	[4, 7, 10, 11]
Infrared camera and sensors	Wavelengths about 1 μm, mounted behind front window	Sensing forward (vehicle and pedestrians)	[4]
Ultrasonic sensors	Distance measurement, facing the front seats	Adapted airbag trigger	
Ultrasonic sensors	Pulse echo and/or triangulation, sensors mounted in bumpers, short range (0.3–3 m)	Parking/shunting aid, near obstacle detection	[12]
Stereo vision sensor	High dynamic range camera (HDRC) in nonlinear CMOS technology	3D object recognition, 2D recognition of traffic signs and lane marking	[13]

of any crash direction via detailed information about crash time, velocity and direction, lane change support, overtake support, etc. Table 2.4-1 lists a variety of existing implementation developments of distance sensors in vehicles. More details of sensor principles and implementations are given in the following sections.

2.4.2 Radar Sensors for Autonomous Intelligent Cruise Control (AICC)

AICC is seen as a first step in the development towards a full collision avoidance system offering $360°$ protection for the vehicle. The majority of automotive manufacturers planned to release their AICC-equipped vehicles from 2000 onwards; Mercedes Benz started in 1999. Radar seems to be the best sensor principle for AICC since alternatives such as optical, infrared and ultrasonic systems may fail under bad weather conditions when they are needed most. Therefore, radar sensors will be the center of interest in the following considerations.

Radar is the abbreviation for radio detecting and ranging, mostly using electromagnetic waves in the microwave region up to about $100\,GHz$, which are also named centimeter waves and millimeter waves because of their wavelengths. The idea of incorporating radar systems into vehicles to improve road traffic safety dates back to the 1970s. Such systems are now reaching the market as recent advances in semiconductor technology have allowed the signal processing requirements, the high angular resolution requirements from physical small antennas, and economic microwave circuitries to be realized.

For distance measurement it is common to use the same antenna for both transmitting and receiving the signal. The received signal power P_r of a distance measurement system as a function of the transmitted power P_t, the relevant antenna area A, reflecting area σ, distance r between the cars, and wavelength λ of microwave radiation is expressed by the radar equation [14]:

$$P_r = \frac{A^2\sigma}{4\pi r^4\lambda^2}\cdot P_t \qquad (1)$$

The received power decreases in proportion to $1/r^4$, and therefore a wide dynamic range has to be covered. With the advent of low-cost integration procedures, such as hybrid and monolithic integration techniques and GaAs foundries, this can be achieved [15].

For distance measurement with microwaves, several radar methods exist that can be divided into two classes: pulse radar methods and continuous wave (CW) radar methods.

Pulse radar systems (see Section 2.4.2.1) generate short microwave pulses, which are emitted by an antenna. The pulses travel through the air to the target, are reflected by the target and move back to the antenna (monostatic device) or back to a second antenna (bistatic device). The time of flight t_0 of the microwave pulses can be measured and transformed into the wanted distance d:

$$d = \frac{t_0}{2c} \qquad (2)$$

(where the velocity of light $c = 2.998\cdot 10^8$ m/s).

Instead of short pulses, in principle long or modulated pulses or pulse trains can also be emitted determining the traveling time using a matched filter. For measuring with sufficient accuracy, however, a major effort to achieve sufficiently short microwave pulses, pulse compression methods, and/or radar signal processing is necessary. Especially the system bandwidth demanded is responsible for high expenditure.

CW radar systems emitt a microwave signal permanently. To obtain distance information these microwave signals must be modulated. The pseudo-noise (PN) code method, for example, applies a pulse series randomly switched by a pseudo-noise generator. The traveling time of this signal is calculated via correlation of the emitted and received signals. Another prominent method is the frequency-modulated continuous wave (FMCW) method, which will be discussed in more detail in Section 2.4.2.2. CW methods reduce the expenditure in the microwave circuitry and shift it partly to digital signal processing algorithms. CW methods also offer better sensitivity than pulse methods.

For automotive applications, different pulse systems as well as CW systems are under development or have already been designed.

2.4.2.1   Microwave Pulse Radar

Pulse radar systems [2] have to measure the time of flight of microwave pulses from sensor to target and back to the sensor again. The traveling time t_0 of a pulse is fairly short because the speed of microwaves equals the speed of light. For example, to distinguish between two neighbouring targets with a distance of 1.5 m, time differences of 10 ns have to be resolved.

The distance resolution of a pulse radar in the case of neighboured targets is limited by the pulse length τ_p. Given two targets T_1 and T_2 with a range difference ΔR (see Figure 2.4-2), the transmitted pulse will be reflected first at T_1 and after a time difference $\Delta t = \Delta R/c$ also at T_2. The path difference between the leading edges of both reflected pulses is $2\Delta R$. Both targets will be separated by the receiver if the distance between the closing edge of the pulse reflected at T_1 and the leading edge of the pulse reflected at T_2 is greater than zero. Therefore, the distance resolution is given by

$$\Delta R_{min} = \frac{1}{2} c\tau_p = \frac{1}{2} c \cdot \frac{1}{B} \tag{3}$$

where $B = 1/\tau_p$ is the bandwidth of the radar system. For monostatic systems the antenna cannot receive while transmitting, hence this equation also gives the minimum measuring distance. A demanded range resolution of 1 m yields a maximum pulse length of 6.7 ns.

The maximum measuring distance, on the other hand, gives the maximum possible pulse repetition frequency, because all reflected pulses must reach the antenna before the next pulse transmission starts. A measuring range of 200 m corresponds to a maximum pulse repetition frequency of 1.5 MHz.

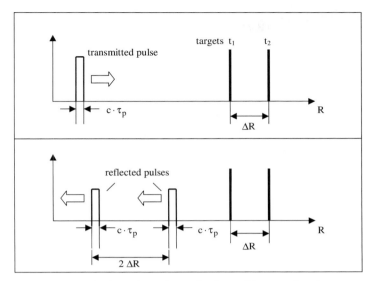

Figure 2.4-2. Distance resolution of a pulse radar in multitarget situations.

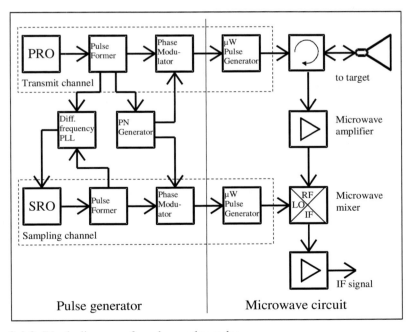

Figure 2.4-3. Block diagram of modern pulse radar.

Figure 2.4-3 shows a block diagram of a modern pulse radar consisting of a transmit channel with a PRO (pulse repetition frequency oscillator), an SRO (sampling repetition frequency oscillator) and a PLL (phase locked loop) for maintaining defined frequency relations between the two channels. This is the pulse generation part of the circuit. In the microwave part there are pulse generators which transform the PRO and SRO signals to coherent microwave pulses, and also microwave mixers and amplifiers. The IF (intermediate frequency) signal further has to be time delayed, sampled and evaluated by digital signal processing. Because of short switching times for the generation of sufficiently short coherent microwave pulses, this type of radar involves expensive microwave circuitry.

2.4.2.2   FMCW Microwave Radar

In the microwave laboratory at the University for Applied Sciences in Aalen we frequently use the FMCW method. Several FMCW systems in the 24 and 94 GHz regions for automotive use and demonstration have been built and tested. The FMCW method has the advantage of comparably low expenditure, because the necessary bandwidth in the microwave circuitry is limited to a small number of microwave components.

The principle of the FMCW method is shown in Figure 2.4-4. The frequency of the transmitted microwave signal f_t increases linearly from the start frequency f_1 to the maximum frequency f_2 within the time interval from 0 to T_w. In the following time interval from T_w to $2T_w$ the frequency decreases again, so we obtain triangle frequency modulation. The received signal f_r shows the same frequency behavior, but delayed for the traveling time t_0 from sensor to the target and back. The received signal also shows a much lower power than the transmitted signal according to the radar equation (Equation (1)). If sufficient radar power is received, the frequency difference between the two microwave signals can be determined. This is the so-called intermediate frequency (IF), which is constant within the time interval from t_0 to T_w and corresponding intervals $T_w + t_0$ to $2T_w$, etc., if the distance to the target does not change. It is also proportional to the time of flight of the signal and proportional to the measuring distance within the interval. The distance d as a function of intermediate frequency f_i according to the short calculation of Figure 2.4-4 yields

$$d = \frac{T_w}{f_2 - f_1} \cdot \frac{c}{2} \cdot f_i \tag{4}$$

A block diagram of the microwave circuitry of a typical monostatic FMCW sensor is shown in Figure 2.4-5. The upper part contains the measuring branch containing the voltage-controlled oscillator (VCO), whose signal is transmitted via the circulator to the antenna and into the air. The circulator passes the received microwave signal to the mixer, to which also part of the oscillator signal is

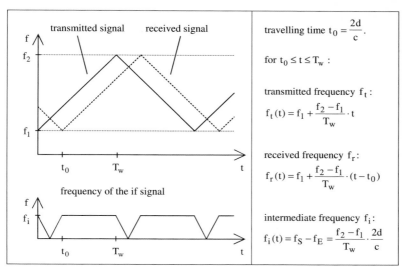

Figure 2.4-4. Principle of the (linear) FMCW method.

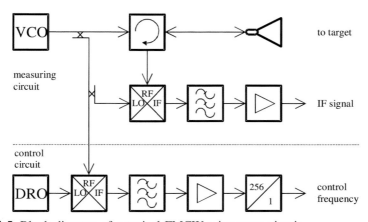

Figure 2.4-5. Block diagram of a typical FMCW microwave circuit.

coupled. The mixer has a nonlinear behavior and produces frequency differences and sums of the input signals. The following component low-pass filters the wanted interference signal in the base band, where it is amplified and sampled.

The modulation bandwidth for automotive applications is usually in the range up to 300 MHz, starting at 76, 60 or 24 GHz. The voltage-frequency behavior of a VCO is normally nonlinear. The overall deviation from linearity can amount to up to 10% and more of the frequency bandwidth. Furthermore, additional ripples may disturb linearity. Therefore, the voltage ramp must be modified to correct these nonlinearities. This calibration could be done in the factory before delivering the system. However, the nonlinearities alter with temperature and time, which decreases the accuracy of the distance sensor the longer it is used.

Therefore, some means of frequency control has to be integrated into the microwave circuitry, eg, via a phase-locked loop and frequency synthesizer. For direct frequency control via counters the microwave frequencies are too high. Another common method is shown in the lower part of Figure 2.4-5. A dielectric resonator oscillator (DRO) gives a stable reference frequency below or above the modulation range of the VCO without any influences of temperature or life time. Mixing the VCO and the DRO signals, the frequency behavior of the VCO can be transformed in the base band and easily counted and calibrated. This can be done between the distance measurements during digital processing of the IF signal of the measuring branch.

The most important device in this circuit is the VCO, eg, a Gunn oscillator. Figure 2.4-6 shows the frequency behavior when modulating the voltage linearly in comparison with the linearized mode. Linear voltage modulation yields nonlinear frequency behavior. The maximum frequency deviation D is more than 12%. Figure 2.4-7 shows the resulting distance errors over distance as a function of D using a broad Dolph-Tschebyscheff window for signal processing and averaging several frequency lines [16]. This algorithm designed for a high precision system gives very small errors in a single target situation using linear frequency modulation with a bandwidth of 1 GHz. Nonlinear modulation gives reasonable errors, which increase with distance.

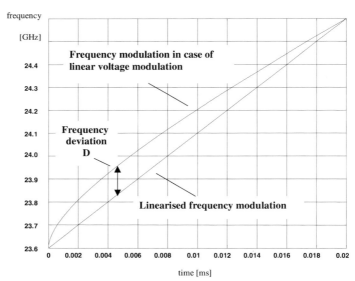

Figure 2.4-6. Frequency behavior with linear voltage modulation in comparison with linearized frequency modulation.

error
[m]

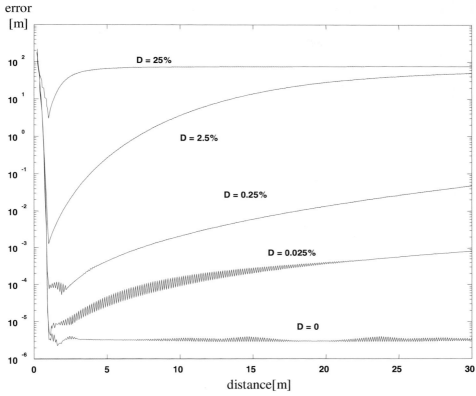

Figure 2.4-7. Example of resulting errors depending on nonlinear modulation for different *D*.

2.4.2.2.1 Digital Signal Processing of FMCW Systems

The IF signal contains distance information by means of the frequency of a harmonic signal. If there were only a single target, the frequency of this signal could be determined by counting the number of zero crossings. However, there can be many targets, cars, trees, buildings, pedestrians, and so on as well as internal reflections, eg, from plugs or antennas. Hence a multifrequent analysis employing fast Fourier transformation (FFT) has to be applied.

The length of the IF signal is limited to the sweep time. It is not allowed to combine several IF signals for FFT because of the arbitrary phase shift. The time-limited IF signal can be interpreted as a not time-limited signal multiplied by a window, eg, a rectangular window. The Fourier transform of a monofrequent harmonic IF signal gives the wanted frequency. However, because multiplying in the time region corresponds to convolution in the frequency region, the transform of the harmonic IF signal is convolved with the transform of the window function, an Si-function in the case of the rectangular window. Because of sampling the time signal, the result also exists only at discrete frequency spacings and gives only a single frequency line, if frequency spacing samples the roots of the Si-function. However, normally there is leakage, there are many frequency lines, and the greatest frequency line does not sample the maximum of the Si-function. Hence this line is only an approximation for the unknown frequency.

The frequency resolution of the FFT is $df = 1/T_w$. Inserting this in Equation (4) yields the corresponding distance spacing of the FMCW distance measurement:

$$\delta r = \frac{T_w}{f_2 - f_1} \cdot \frac{c}{2} \cdot df = \frac{c}{2B} \tag{5}$$

This corresponds to Equation (3). Thus again the bandwidth $B = f_2 - f_1$ of the system defines the distance resolution.

The frequency resolution of the FFT in principle can be improved by increasing the measuring time. Because of the limited bandwidth of the microwave module, the sweep time also has to be increased, without changing the start and stop frequencies. Hence the distance resolution does not change (see Equation (5)). To distinguish between two targets, their distance must be at least as great as the distance spacing of FFT. Hence here FMCW radar has the same limitations as pulse radar.

Determining the position of a target according to Equation (4) gives a distance error in the range of one to two times the distance spacing, which is about 0.5–0.8 m for long-range measurement (AICC). In general this is not good enough. To calculate the exact position of a single target we use the information of the height of the neighbouring frequency lines of the maximum, ie, we make an interpolation. There are many possibilities for doing this.

2.4.2.2.2 Interpolation Methods in the Frequency Domain

2.4.2.2.2.1 Center of Gravity of Two Frequency Lines

Using the two highest frequency lines in a power spectrum the wanted frequency can be determined easily as the center of gravity of the two lines, if there is no or only little noise (Figure 2.4-8). The FFT of the rectangular window is (N = number of samples, Ω = normalized frequency according to Shannon):

$$W_N(\Omega) = \frac{\sin(N\Omega/2)}{\sin(\Omega/2)} , \quad 0 \leq \Omega \leq 2\pi \tag{6}$$

frequency belonging to the maximum of the continuous spectrum, wanted frequency (k number of frequency line):

$$\Omega_0 = (k + \Delta k)\frac{2\pi}{N} \tag{7}$$

neighbouring frequency lines:

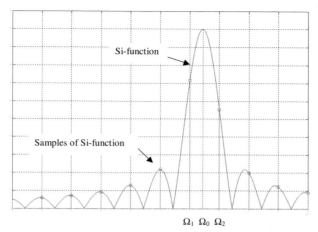

Ω₁ Ω₀ Ω₂

Figure 2.4-8. Leakage and center of gravity of two frequencies.

$$\Omega_1 = k\frac{2\pi}{N} \qquad \Omega_2 = (k+1)\frac{2\pi}{N} \tag{8}$$

center of gravity of the neighbouring frequency lines:

$$\frac{W_N(\Omega_1)\Omega_1 + W_N(\Omega_2)\Omega_2}{W_N(\Omega_1) + W_N(\Omega_2)} = \frac{\dfrac{\sin(-\pi\Delta k)}{-\pi\Delta k/N}\cdot\dfrac{2\pi k}{N} + \dfrac{\sin(\pi\Delta k)}{\pi(1-\Delta k)/N}\cdot\dfrac{2\pi(k+1)}{N}}{\dfrac{\sin(-\pi\Delta k)}{-\pi\Delta k/N} + \dfrac{\sin[\pi(1-\Delta k)]}{\pi(1-\Delta k)/N}} = \Omega_0 \tag{9}$$

where

$$\sin\left(\frac{\pi\Delta k}{N}\right) \approx \frac{\pi\Delta k}{N} \tag{10}$$

if the number of samples N is sufficiently large. Hence the wanted frequency can be determined exactly. However, the heights of two neighbouring lines can differ substantially, so noise in the system can cause large errors.

2.4.2.2.2.2 Zero-padding

Zero-padding means adding of zeros to the sampled IF data. Hence the measuring time can be increased by a factor of n:

$$T_o \Rightarrow nT_0 \tag{11}$$

This yields an improved frequency resolution (the Si-function is sampled n times more often):

$$F_0 \Rightarrow \frac{F_0}{n} \tag{12}$$

and the distance error decreases also by the factor n. However, this method requires also increased memory space by the same factor and increased computing time.

2.4.2.2.2.3 Rotation of the Nyquist Spacing

If the frequency spacing of the FFT or DFT (discrete Fourier transform) does not fit to the wanted frequency (see Figure 2.4-7) it can be shifted ('rotated') by modifying the transform to

$$Y(k + \Delta k) = \sum_{i=0}^{N-1} y(i) e^{-j2\pi i(k+\Delta k)/N} = Y(k + \Delta k) = \sum_{i=0}^{N-1} [y(i) e^{-j2\pi i\Delta k/N}] e^{-j2\pi ik/N}$$

$$\tag{13}$$

The standard FFT can be used for this by interchanging the data y by 'rotated' data y':

$$y'(i) = y(i) \cdot e^{-j2\pi i\Delta k/N} \quad \text{(rotation operator)} \tag{14}$$

The optimal value of Δk has to be determined by a gradient search method. As a criterion for success, maximizing the spectral line or minimizing the sum of the other lines can be used. This search method also needs substantial computing time.

2.4.2.2.2.4 Center of Gravity Combined with Other Windows

The noise influence in the previous method is decreased by using a Hamming window instead of the rectangular window. Hence the main peak of the transformed window function becomes broader. An optimum for frequency interpolation in this case is the center of gravity of four frequency lines.

The use of a Dolph-Tschebyscheff window is good for monofrequency signals with little noise. Here the center of gravity of 13 lines seems optimal [16].

Other window functions require other averaging parameters according to the width of their main peak.

2.4.2.2.2.5 Hilbert Transform

The Hilbert transform treats the real measured signal as an analytic signal. Hence the real and imaginary parts of the resulting complex function contain the same frequencies but with a phase shift of 90° [17]. Starting with a frequency mixture with amplitudes A_i:

$$x(t) = \sum_i A_i \sin(2\pi f_i t + \varphi_i) \tag{15}$$

and applying FFT, filtering in the frequency domain, and inverse FFT, we obtain the monofrequency complex time signal:

$$x_{\text{real}} = A_1 \sin(2\pi f_1 t + \varphi_1) \qquad x_{\text{imag}} = A_1 \cos(2\pi f_1 t + \varphi_1) \tag{16}$$

This signal simply allows the calculation of the phase and the wanted frequency:

$$\varphi = \arctan\left(\frac{x_{\text{imag}}}{x_{\text{real}}}\right) = 2\pi f_1 t \tag{17}$$

$$f_1 = \frac{\varphi(t_2) - \varphi(t_1)}{2\pi(t_2 - t_1)} \tag{18}$$

The processing time for this method is fairly long. However, it gives the advantage of yielding the phase function over time, which can be used to control the linearity of the frequency modulation. Hence during operation online any deviation from linearity can be detected and corrected [18].

2.4.2.2.2.6 Parabolic Algorithm (Hamming Window)

Using the Hamming window, the main peak of the Si-function becomes broader, so the frequency transform of the Hamming window will be sampled at three points within the main peak. Its shape can be approximated fairly well by a parabola. The three samples serve to calculate the three parameters describing a parabola (curvature a and coordinates of the extremum f_s and y_s) by solving a system of three linear equations:

$$\begin{aligned} y_1 &= a(f_1 - f_s)^2 + y_s \\ y_2 &= a(f_2 - f_s)^2 + y_s \\ y_3 &= a(f_3 - f_s)^2 + y_s \end{aligned} \tag{19}$$

Combining this method with zero-padding gives even more samples within the main lobe of the spectrum. In this case a least-squares fit is advantageous for the calculation.

Zero-padding also improves the frequency resolution. The frequency spacing from the highest frequency line to two neighbouring lines decreases. Using the parabolic algorithm with only three samples in this case gives very good results with minimal calculation expenditure. Therefore, we favor this method [16]. The resulting distance error in monotarget situations can thus be reduced to a fraction of the distance spacing.

2.4.2.2.3 Elimination of the Doppler Shift

Measuring distances of moving targets via the FMCW method means analyzing the frequency of the reflected signal. According to the Doppler effect an additional frequency shift, the Doppler shift F_d, superimposes, which is proportional to the microwave frequency f and the velocity v of the target:

$$F_d = f \cdot \frac{v}{c} \qquad (20)$$

Depending on the design of the FMCW system parameters, this Doppler frequency can disturb the distance measurement substantially. In this case we combine the result for the intermediate frequency f_i of the rising frequency ramp with the intermediate frequency of the falling ramp. In the first case the Doppler shift is added to the intermediate frequency and in the second case it is subtracted. Hence both Doppler shift and intermediate frequency can be determined.

2.4.2.3 Frequency Shift Keying (FSK) Doppler Radar

A special type of Doppler radar system with few bandwidth requirements can be used to track not only the velocities of multiple targets but also range and range rate for each target. The system uses FMCW radar transmission with frequency shift keying [19].

The type of FMCW/FSK frequency modulation is shown in Figure 2.4-9. A minimum of two different frequencies must be used in order to measure the range. Additional transmit frequencies can give better data verification and improved tracking algorithms at additional system cost. In this example, the radar transmitter frequency is periodically shifted about 500 kHz at a rate of 100 kHz.

Because of the Doppler effect (Equation (20)) at 24.125 GHz, for example, we get 45 Hz per km/h target closing rate. If the system covers a closing rate range from 0.16 to 160 km/h, the Doppler frequency band extends from 7 to 7200 Hz. A frequency response spectrum is shown in Figure 2.4-10. The vertical axis represents signal strength and the horizontal axis represents Doppler frequency or target closing rate. Multiple targets are distinguished by their different Doppler frequencies.

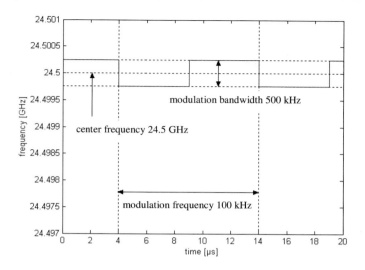

Figure 2.4-9. Frequency modulation of FSK Doppler radar.

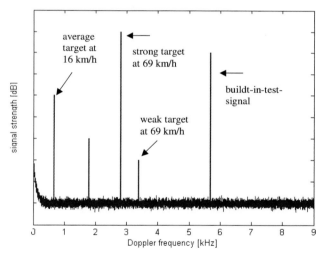

Figure 2.4-10. Frequency response spectrum of FSK Doppler radar.

The digital signal processing produces a frequency output spectrum like Figure 2.4-10 for each of the two transmitted frequencies. A comparison of Doppler phase angles between two radar frequencies for each target will be proportional to the round trip signal travel time to that target and thus generate the range to the target. The sign of the closing rate, indicating whether it is closing or opening, is determined by the lead/lag relationship of the two transmit frequencies.

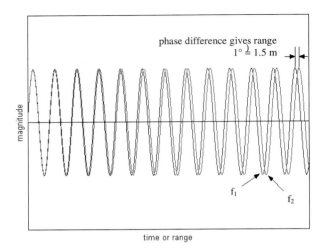

Figure 2.4-11. Range determination via phase difference.

Figure 2.4-11 shows how the range is determined by measuring the phase difference between the signals received on the two transmit frequencies. The slight difference in the two transmit frequencies results in two signals whose phase difference is proportional to the round trip signal path time period. For a 24 GHz system, for example, range is related to phase difference at about 1.5 m per phase degree. The maximum range that can be measured with this technique is equivalent to the range equivalent of a 180° phase shift.

The FSK method can be used only for moving targets with different velocities. Besides this limitation it is advantageous because of the comparably small system bandwidth [6, 19, 20].

2.4.2.4   Antenna Design and Radar Technology

For use in automobiles, any sensor should be limited in size and fit into the body of the car without disturbing the car design. On the other hand, microwave antennas for long-range detection need a narrow beamwidth for generating high gain and object resolution. The half-power beam width β in degrees is given by wavelength λ and antenna aperture A [14]:

$$\beta = 51° \cdot \frac{c}{Af} = 51° \cdot \frac{\lambda}{A} \tag{21}$$

Hence a beam width of 1° in a 24 GHz radar system would require an antenna aperture of 64 cm, which is much too large to be acceptable. Therefore, AICC radar has to be realized with comparably high microwave frequencies of 60 GHz

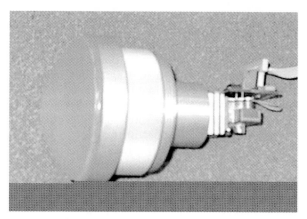

Figure 2.4-12. Experimental 24 GHz FMCW system with dielectric lens antenna.

(Japan) or 76 GHz (Europe). Real systems with a 3° beam width at 76 GHz require a minimum antenna aperture of about 10 cm. Using radome they can be easily integrated into any car design, secure against mechanical damage and rain, etc. In the future even higher microwave frequencies may be chosen, possibly up to 152 GHz and beyond to reduce sensor size.

To distinguish between preceding cars in different lanes or obstacles beside the road, especially on curves, a single static antenna beam is not sufficient. Therefore, the antenna beam should scan the scene. This could be achieved by mechanically moving the antenna. Much more elegant is electrical switching be-

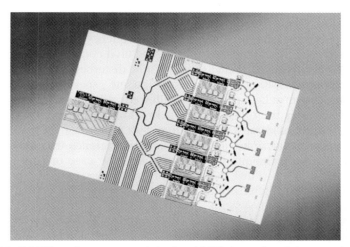

Figure 2.4-13. 76 GHz microwave module for AICC (reproduced by permission of Siemens).

tween several different antenna feeds or even electronic phase switching of the characteristic of planar antenna arrays.

Dishes and lens antennas give the cheapest realizations. Figure 2.4-12 shows an experimental 24 GHz FMCW system with dielectric lens antenna used in our laboratory for distance measurement up to 100 m. For measuring in three directions to give the AICC directivity information, DASA [21] developed a horn antenna with dielectric lens working at 76 GHz. The antenna length is 70 mm, aperture 140 mm, beam width 2° in azimuth and elevation with sidelobes less than 15 dB. The determination of angle with respect to the car axis is done by comparing the complex amplitudes or the radar reflections of the three beams [6]. Flat phase array antennas are realized mostly in the frequency range up to 24 GHz, and in the future they will also extend to 76 GHz.

The central and most critical component in millimeter wave radar is the oscillator. Gunn diodes in waveguide cavities are still the most efficient but are costly and will be replaced by monolithic microwave integrated circuits (MMICs) [22, 23]. Using waveguide components the costs increase not only linearly with frequency but to the second power or even higher. Therefore, integration of all microwave components into finline or microstrip circuits is aimed at (Figure 2.4-13). The split-block technique already allows planar hybrid microstrip integration at comparably, but for automotive not sufficient, low cost. Other than in micro systems technology, whose boom is based on silicon technology, miniaturizing millimeter wave technology needs progress in III/V-semiconductor technology (GaAs and InP). In this field, progress via realization of MMICs will replace the current hybrid systems. A fully integrated MMIC for 76 GHz was presented by United Monolithic Semiconductors in 1999 [24].

2.4.2.5   General System Considerations

Microwave applications are not allowed in any frequency band because of possibly harmful interferences. For example, military and civil communication systems also employ microwave frequencies and should not be disturbed. For industrial applications, mobile communication, satellite television, etc., special frequency bands are reserved. Automotive radar in Europe is allowed in the 24 GHz and 76 GHz regions, and Japan uses the 60 GHz region. In addition, the microwave circuits in different cars on the road must not interfere with each other. FCC regulations demand that a radar device must not accept any interference that may cause undesired operation. One technique to achieve this in the case of cars is to polarize the microwaves to 45°, so that the waves of facing cars have 90° polarization to each other and can be filtered. Other possibilities employ PN coding of the transmitted waves or use radar systems with bandwidths as small as possible (FSK). There is electromagnetic compatibility (EMC) and radio type approval legislation that all radar systems must meet before they can be operated and sold legally within Europe [15]. The European standards of technical characteristics and test methods are defined in EN 301 091 for 76–77 GHz radar. National regulations have to be harmonized within all European countries. Similar regulations exist in the USA (FCC) and Japan, where a certificate of conformity to technical standards from the Radio Equipment Inspection and Certification Institute must be obtained. Other implications could be biological effects of electrosmog on humans. Research on these themes is increasingly carried out, as the public cannot accept neglect of biological effects. The possible effects of 76 GHz AICC radar on humans was the subject of a recent study, but no significant effect was found [25].

2.4.3 Infrared Scanning and Other Imaging Systems

Infrared sensors are available as low-cost systems. For AICC an infrared (IR) sensor system has been designed as a pulsed multibeam scanning system [4, 10]. Such systems cover up to about 120 m in clear weather. The wavelength of about 1 μm gives better performance than the human eye in bad weather. For scanning of the scene the system uses a mechanical rotation unit about the vertical axis, driven by a motor belt assembly at a speed of 10 Hz. This yields a scanning angle of 180°. The beam divergence is 1° vertical and 2° horizontal, and the dynamic range is 60 dB. Other systems even apply two-dimensional scanning methods [11]. These infrared laser radar systems can serve for both front and rear obstacle detection and for determining the position of the car regarding the white lines on the road. For distance measurement the triangulation method is used for distances under 30 m and time of flight measurement for longer distances. Owing to its reduced dimensions, the infrared laser radar detection systems can be carried on-board the vehicle, eg, in the headlights.

Methods of video picture processing also can serve for determining distances, recognizing lanes and lane departure warnings. Charge-coupled device (CCD) cameras are already used for sensing the white lines on the road looking forward through the front window. Looking back, CCD cameras serve for sensing rearward vehicles. Miniaturized systems perform object detection and measuring by stereo vision [13].

An attractive prospect for future solutions concerns radar imaging. A first step has already been made using an improved AICC radar system with increased resolution [9]. In high-resolution images of range vs. azimuth, vehicles appear as laterally and longitudinally extended objects. Caused by reflections from the road surface, the complete bottom sides can be seen, allowing for the recognition of vehicles and distinction of different classes (cars, vans, trucks). Even obscured vehicles become visible, which provides very important information for stop and go traffic. An azimuth resolution of about 1° and a range resolution of about 1 m proved to be sufficient.

Further perspectives involve the integration of several sensor principles. The IR laser sensor could be complemented with a passive pyroelectric IR detector and a far-looking microwave radar. The potential for combining different types of sensors into one system is to make the most of the advantages that each offers with respect to detection capability and resistance to environmental effects and thus provide better support for enhancing vehicle safety.

2.4.4   Short-range Distance Measurement as a Parking Aid

Today's short-range distance measurement is based on ultrasonic technology. It is accepted as a useful support in low-speed parking maneuvers. The technical design of adequate systems is again determined by a variety of different requirements such as cost, performance, and esthetics [26]. Statistics show that about 40% of accidents involving trucks happen when reversing. Cars with modern aerodynamic styling offer little sight to the rear of the car. Therefore, several ultrasonic sensors are mounted into the rear bumper to cover close rear obstacles. Triangulation with at a minimum two sensors serve for distance measurement. The range is 1.6 m for cars and 3 m for trucks. Distance warning is given via optical signals and, more important, acoustically by increasing beep frequency. The measuring errors amount to 0–20 cm. Below a 30 cm distance obstacles are recognized but without distance resolution.

Ultrasonic sensors must have direct contact with the air and therefore cannot be hidden behind the bumper, etc. This is a constraint for the design of a car. An alternative to ultrasonic parking is radar, if it can compete concerning costs. Radar gives the designer more freedom and is not dependent on environmental conditions. Because of the wide angle for near-obstacle detection, comparably low microwave frequencies are possible, which helps in reducing costs. On the other hand, distance resolution has to be increased compared with AICC radar, which requires much larger frequency bandwidths and increases costs. Figure 2.4-14 shows a 24 GHz system designed as a parking aid.

Figure 2.4-14. 24 GHz microwave module for a parking aid (reproduced by permission of Siemens).

2.4.5 Conclusion

Dealing with sensors for distance measurement and with their applications in automobiles is and will remain a very interesting and challenging task. We are currently observing a breakthrough in radar technology which, together with the advances in signal processing technology, allows new driver assistance systems like AICC. Thus the driver of a car has the benefit of greater safety and comfort. Because of increasing traffic density this may be complemented by automatic driving in convoys, where video or infrared laser scanning may form the base for autonomous driving within a lane. This may be complemented by the development of surround sensing systems for collision avoidance in all directions.

2.4.6 References

[1] Bannatyne, R., in: *Automotive Handbook;* New York: McGraw-Hill, 2000, p. 29.
[2] Winner, H., in: *Automotive Handbook*; New York: McGraw-Hill, 2000, p. 30.
[3] Basten, M., Leo, G., *Konferenzbericht Fahrzeug- und Motorentechnik, 7. Aachener Kolloquium, Aachen*; 1998, Vol. 2, pp. 1311–1329.
[4] 'Mazda's second generation prototype advanced safety vehicle unvealed', *Asia-Pacific Automotive Report*; 1999, No. 309, pp. 33–34.
[5] Uhler, W., Weilkes, M., *StopGo: System Design and Functionality of an extended ACC*, presented at the Konferenzbericht Fahrzeug- und Motorentechnik, 9th Aachener Kolloquium, Aachen, 2000.
[6] Olbrich, H., Beez, T., Lucas, B., Mayer, H., Winter, K., presented at the Congress advanced Controls and Vehicle Navigation Systems, Detroit, 23–26 February, 1998, SP-1332.
[7] 'Nissan begins launch of microwave adaptive cruise control', *Asia-Pacific Automotive Report*; 1999, No. 302, pp. 21–23.
[8] Redfern, S., Oxley, C., Dawson, D., Bird, J., Hilder, G., Prime, B., Brown, T., Spencer, D., *Microwave Eng. Eur.* (1999/2000) 29–33.
[9] Schneider, R., Wenger, J., presented at the 1999 IEEE MTT-S, Anaheim, CA, 13–19 June 1999, paper MO3E-3.
[10] Najami, A., Mahrane, A., Vialaret, G., Esteve, D., *Sen. Actuators A* **41/42** (1994) 47–52.
[11] Osugi, K., Miyauchi, K., in: *4th International Symposium on Advanced Vehicle Control, AVEC*; 1998, pp. 735–740.
[12] Hötzel, J., Knoll, P., Noll, M., Weber, J., *VDI-Berichte* **1188** (1995) 333–339.
[13] Seger, U., Knoll, P.M., Stiller, C., *Sensor Vision and Collision Warning Systems*; Las Vegas: Society of Automotive Engineers, 2000-01-C001.
[14] Detlefsen, J., *Radartechnik;* Berlin: Springer, 1989.
[15] Lawton, S., Andrews, C., Topham, D., *J. Navig.* **53** (2000) 48–53.
[16] Otto, J., *Methoden zur Frequenzbestimmung bei mono- und multifrequenten Signalen*; Wiesbaden: Tagungsband MessComp, 1995, pp. 113–117.

[17] Otto, J., *Digitale Signalverarbeitung bei der Mikrowellen-Füllstandsmessung*; München: Tagungsband DSP Deutschland, 1996, pp. 65–74.

[18] Otto, J., *Selbstkalibration eines FMCW-Abstandsmeßgeräts mittels digitaler Phasenanalyse*; Wiesbaden: Tagungsband MessComp, 1995, pp. 143–147.

[19] Woll, J.D., in: *Intelligent Vehicles 95,* Detroit, 25–26 September; 1995, IEEE: Piscataway, NJ, pp. 42–47.

[20] Grosch, T., *Proc., SPIE* **2463** (1995) 239–247.

[21] Huder, B., *ITG Fachtag. Dresden* **128** (1994) 37–42.

[22] Meinel, H., in: *28th EuMC 1998,* Amsterdam, 5–9 September, 1998; Tutorial MT-E8, pp. 619–641.

[23] Hager, W., Schroth, J., Meinel, H., in: *Printed Circuits*, Waiblingen, 28–29 Nov., 1996, pp. 7.1–7.23.

[24] Caminade, M., Domnesque, D., Alleaume, P., Mallet, A., Pons, D., Daembkes, H. presented at the 1999 IEEE MTT-S, Anaheim, CA, 13–19 June, 1999, paper TH2A-2.

[25] Waldmann, J., Landstorfer, F., *Untersuchung möglicher Wirkungen eines 77 GHz-Kfz-Abstandsradars auf das vegetative Nervensystem des Menschen*; Stuttgart: Institut für Hochfrequenztechnik, Universität Stuttgart, 2000.

[26] Hötzel, J., Knoll, P., Noll, M., Weber, J., *VDI-Bericht* **1188** (1995) 333–339.

List of Symbols and Abbreviations

Symbol	Designation
a	curvature
A	antenna area/aperture
B	bandwidth
c	velocity of light
d	distance
D	frequency deviation
f	microwave frequency
f_1	start frequency
f_2	maximum frequency
f_i	intermediate frequency
f_r	received signal
f_s	coordinate of extremum
f_t	transmitted signal
F_0	frequency resolution
F_d	Doppler shift
n	factor increase
N	number of samples
P_r	received signal power
P_t	transmitted power
r	distance between cars
ΔR	range difference
t_0	time of flight/traveling time

Symbol	Designation
Δt	time difference
a	curvature
T	measuring time
T_{w}	wobble time
x	complex time signal
y	data
y'	rotated data
y_{s}	coordinate of extremum
v	velocity of target
β	half-power beam width
φ	phase
λ	wavelength of microwave radiation
Ω	normalized frequency
σ	reflecting area
τ_{p}	pulse length

Abbreviation	Explanation
ACC	automatic cruise control
AICC	autonomous intelligent cruise control
CCD	charge-coupled device
CMOS	complementary metal oxide semiconductor
CW	continuous wave
DRO	dielectric resonator oscillator
DFT	discrete Fourier transformation
EMC	electromagnetic compatibility
FCC	federal communications commission
FFT	fast Fourier transformation
FMCW	frequency-modulated continuous wave
FSK	frequency shift keying
HDRC	high dynamic range camera
IF	intermediate frequency
MMIC	monolithic microwave integrated circuit
PLL	phase locked loop
PN	pseudo-noise
PRO	pulse-repetition frequency oscillator
Radar	radio detecting and ranging
SRO	sampling repetition frequency oscillator
VCO	voltage-controlled oscillator

2.5 Micropropulsion for Space – A Survey of MEMS-based Micro Thrusters and their Solid Propellant Technology

C. Rossi, CNRS Researcher, France

Abstract

In the last two decades, advances in microsystems technology have made micro mechanics mature enough to envisage miniaturization in the space industry. The motivation behind this development is the reductions in mass, satellite costs and launch costs in parallel with an increase in the reliability and flexibility of satellites. A cluster of many small satellites, commanded from a mothership, or totally independent, will, without any doubt, reduce the risk of mission failure and increase mission flexibility. This revolution in the space industry will rely on propulsion system development to ensure the maneuvering and fine positioning of micro and nano satellites (defined as satellites with mass between 20–100 kg and under 20 kg, respectively). This paper is provides a state-of-the-art review of micro thruster technologies for space applications. the intention is to overview the different technical solutions under investigation in the micropropulsion field, to assess their merits and disadvantages and to try to identify the promising ones in relation to future missions needs. First, the principles of propulsion and basic concepts and equations useful in describing and comparing propulsion systems are described. Then, a review micropropulsion needs for space is made and the options and technologies available for propulsion are presented. An outline is given of the technological efforts made in miniaturizing propulsion systems through examples from current research programs and the different micropropulsion technologies are compared. After illustrating the assets of microsystem technology in the micropropulsion field by the detailed presentation of one particular micropropulsion option investigated at LAAS-CNRS, namely solid propellant micro thruster arrays, a short discussion is given of the capability of microsystem technology to serve micropropulsion needs.

Keywords: micro satellite; micropropulsion; micro thruster; thrust; microsystem technology; silicon

Contents

2.5.1   Introduction

The most significant benefits of using microsystem technology in space applications is the reduction in size of parts and connections because of better integration of each component, a decrease in material variability and increased reliability. This will result in mass reduction and in lower fabrication costs. Beyond the integration of microsystem components, the construction of very small satellites using microsystem technology will become a reality in the next decade, which will constitute a revolution in space industry: (1) on the one hand in the fabrication of the satellites, and (2) on the other hand in the manner in which they will be used. There is growing interest within the space community in replacing some large satellites by clusters of small satellites.

Building a cluster of small satellites should be cheaper, more robust and more versatile than building a single, huge satellite. If one fails another could replace it and take the same function. Typical science missions well adapted for micro spacecraft application are space science applications, asteroid missions, and multi-spacecraft observer clusters. These clusters could be launched from either a large satellite (the called mother-ship) or from a launch vehicle. As far as we can assess at present, micro and nano spacecraft concepts may be of interest in already identified scientific missions and may also open up space to new missions. For example, distributed nanosatellite clusters could be launched (1) for disaster and magnetospheric monitoring, (2) for a global survey of the planet, and (3) for the inspection of other spacecraft, etc. All cases heavily rely on the propulsion capability for maneuvering. Propulsion is a key point in the miniaturization of spacecraft because micro and nanospacecraft would need very small (below the newton scale) and very accurate force to achieve stabilization, pointing and station keeping. The required degree of force impulse precision, level of thrust [1] weight and space requirements cannot be reached with conventional propulsion systems. Therefore, in the last decade, micropropulsion has been an active field of research and in numerous research programs the technological possibilities for miniaturizing conventional thrusters have been investigated [2–7]. Our intention with this survey is to overview the different technical solutions under investigation in the micropropulsion field, assess their merits and disadvantages and try to identify the promising ones in relation to future missions needs.

The chapter is divided into six main sections. In Section 2.5.2, we present briefly the principle of propulsion and basic concepts and equations useful in describing and comparing propulsion systems. In Section 2.5.3, we review the micropropulsion needs for space and present the options and technologies available for propulsion and those which can be miniaturized. In Section 2.5.4, we outline the technological efforts made in miniaturizing propulsion systems through examples from current research programs. We compare the different micropropulsion technology on the basis of three criteria: thrust performance, specific power, and technological complexity.

In Section 2.5.5, we illustrate the assets of microsystem technology in the micropropulsion field by the detailed presentation of one low-cost and low-consumption micropropulsion option: the solid propellant technology. Section 2.5.6 concludes the survey with a short discussion on the capability of microsystem technology in making miniaturized, precise, and low-cost devices.

2.5.2   Principles of Propulsion and Basic Relations

The basic parameters for a propulsion device are the thrust F, the total impulse I_t or impulse bit, the minimum impulse I_m and specific impulse I_{sp} (Figure 2.5.1).

2.5.2.1   Calculation of the Thrust Force

The thrust force F of a particular thruster can be determined from the mass flow rate of the propellant times the exhaust velocity V_e, added to the thrust due to the pressure against the exhaust nozzle with area A_e:

$$F = \dot{m}V_e + (P_{ext} - P_e) \times A_e \, . \tag{2.5.1}$$

Equation 2.5.1 is valid assuming a uniform axial velocity that does not vary across the flow area.

2.5.2.2   Calculation of the Specific Impulse and Total Impulse

The specific impulse I_{sp} is defined as the thrust F divided by the mass flow rate $\dot{m}$ of propellant through the thruster, and is a function of propellant and thruster types:

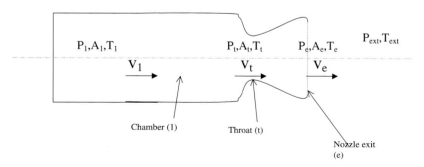

Figure 2.5.1. Gas characteristics in a chamber and nozzle.

$$I_{sp} = \frac{F}{g_0 \dot{m}} \qquad (2.5.2)$$

where g_0 is the gravitational acceleration.

Equation (2.5.2) shows that a higher I_{sp} means that less propellant mass will be consumed for a given mission, so higher is generally better.

For a chemical thruster, the theoretical specific impulse is given by the relation

$$I_{sp} \cong \frac{1}{g_0} \sqrt{\frac{2k}{k-1} \times \frac{R}{M} T_1 \left[1 - \left(\frac{P_e}{P_1} \right)^{\frac{k-1}{k}} \right]} \qquad (2.5.3)$$

with P_e is the pressure at the exit section, P_1 and T_1 are the stagnation condition pressure and temperature, respectively, M is molecular weight of the exhaust gas, R is the universal gas constant, and k is the ratio of specific heats (C_p/C_v).

The total impulse is defined as the time integral of thrust:

$$I_t = \int F \mathrm{d}t = \int I_{sp} \, g_0 \, \dot{m} \, \mathrm{d}t . \qquad (2.5.4)$$

2.5.3   Survey of Propulsion Needs and Options

During its trajectory, a satellite experiences various disturbances in its velocity and stability due to forces and torques that tend to translate and rotate the satellite. Perturbations inducing forces can be categorized.

2.5.3.1   Forces Acting on Spacecraft

The external forces commonly acting on a spacecraft are aerodynamic forces and gravitational attraction. Other forces such as wind and solar radiation pressure are small and generally can be neglected for first calculations. These disturbances slow the satellite and destabilize it.

1. The aerodynamic drag. This factor is most significant at low orbital altitude. The drag F_d in a direction opposite to the velocity is due to the resistance of the satellite to motion in a fluid. It is expressed as a function of the flight speed, V, the density of the surrounding fluid, ρ, and a typical surface area, A:

$$F_d = C_d \times \frac{1}{2} \rho A V^2$$

Table 2.5.1. Air density as a function of altitude

Altitude (km)	Air density, ρ (kg/m^3)	Altitude (km)	Air density, ρ (kg/m^3)	Altitude (km)	Air density, ρ (kg/m^3)
0	1.22	25	4.01×10^{-2}	160	1.23×10^{-9}
1	1.1117	50	1.03×10^{-3}	200	2.54×10^{-10}
3	9.09×10^{-1}	75	3.49×10^{-5}	400	2.80×10^{-12}
5	7.63×10^{-1}	100	5.60×10^{-7}	600	2.137×10^{-13}
10	4.14×10^{-1}	130	8.15×10^{-9}	1000	3.561×10^{-15}

where C_d is the drag coefficient (~ 2). The density of air at different altitudes [8] is reported in Table 2.5.1.

Commonly, a ballistic coefficient W is introduced to parameterize the orbit decay rates. W is defined as the satellite mass M_S divided by the typical surface area A times the drag coefficient C_d:

$$W = \frac{M_S}{A C_d} \, .$$

In Figure 2.5.2 are presented the orbit decay rates as a function of the ballistic coefficient for four orbit altitudes. It illustrates that above altitude around 1000 km the aerodynamic drag becomes negligible.

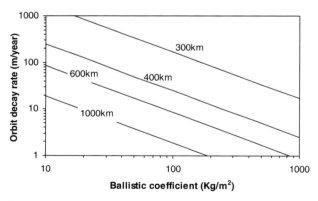

Figure 2.5.2. Orbit decay as a function of the ballistic coefficient W and altitude.

2. Solar radiation. Solar radiation dominates at high altitudes (above 800 km) and is due to the bombardment of the solar photons on the satellite surface. The solar radiation pressure (N/m^2) is given by the Equation (2.5.8)

$$p = 4.5 \times 10^{-6} \times \cos \varphi [(1 - k_s) \cos \varphi + 0.67 k_d] \tag{2.5.5}$$

where φ is the angle between the incident direction and the normal of the im-
pacted surface, and k_s and k_d are the specular and diffuse coefficients of reflectiv-
ity with typical values of 0.9 and 0.5, respectively.

The torque generated on the craft because of the solar radiation is expressed
as

$$T = p \times A \times L \qquad (2.5.6)$$

where A is the area and L is the distance between the spacecraft center of mass
and the center of solar pressure.

3. Gravity gradients. Gravitational torque in a spacecraft results from variation
of the gravitational force. The determination of this torque requires a knowledge
of the gravitational field distribution. This effect is obviously significant in large
and heavy spacecraft operating at low altitude.

4. The earth's magnetic field producing a magnetic torque, T_p. The earth's
magnetic field always fluctuates in direction and intensity because of the magnetic storm effect and others. It decreases with the orbital altitude R as $1/R^3$. T_p
ranges from 10^{-6} to 10^{-4} N m at low orbital altitude.

5. Spacecraft operations such as solar panel deployment and orbital changing.
These also induce perturbation torque.

2.5.3.2 Propulsion Requirements

A propulsion module is required to compensate for the forces acting on the satellite, for orbital maneuvering and for satellite attitude control. Propulsion requirements are commonly expressed in terms of the velocity increment (ΔV) necessary to compensate for the orbital disturbances or to realize an orbital change.
From the ΔV requirements calculated from the mission specifications and spacecraft characteristics, the propulsion requirement is expressed in terms of propellant need or consumption. Of course, it depends on the choice of the propulsion
system. Two cases are usually distinguished: the propellant needs for orbital control and the propellant needs for attitude control.

2.5.3.2.1 Propellant Mass Needed to Realize a ΔV (Orbital Control)

The thrust required to make a change in the spacecraft velocity, ΔV, is function
of the mass of the spacecraft, M_S:

$$F = M_S a = M_S \left(\frac{dV}{dt} \right) \qquad (2.5.7)$$

where a is the spacecraft acceleration. Since the change in velocity is accomplished by consumption of propellant, the total mass M_S also changes:

$$F(\mathrm{d}t) = M_S(\mathrm{d}V) + V(\mathrm{d}M_S) \ . \tag{2.5.8}$$

To simplify, we assume that the amount of propellant used during the change of velocity maneuver is very small compared with the total mass of the spacecraft. Then the thrust needed for a fixed ΔV required in a fixed amount of time is about

$$\int F\mathrm{d}t = M_S\Delta V \ . \tag{2.5.9}$$

In reality, as the total mass of the spacecraft decreases during a maneuver, the required total impulse is reduced. $M_S\Delta V$ represents also the total impulse defined in Equation (2.5.4).

From Equations (2.5.2) and (2.5.7), we have

$$\int F\mathrm{d}t = \int I_{sp} \, g_0 \, \dot{m} \, \mathrm{d}t = M_S\Delta V \quad \text{and} \quad \int \dot{m} \, \mathrm{d}t = m_p \ .$$

Finally, the propellant mass needed to increment the spacecraft velocity is given by the relation

$$m_p = \frac{M_S\Delta V}{I_{sp} \, g_0} \ . \tag{2.5.10}$$

The propellant mass fraction is often used as a criterion of comparison to evaluate and compare the propulsion systems. The propellant mass fraction η is

$$\eta = \frac{m_p}{M_S + m_p} \ . \tag{2.5.11}$$

The value of the propellant mass fraction indicates the quality of the design. A high value of η is of course desirable.

2.5.3.2.2 Propellant Mass Needed to Realize an Angular Rotation (Attitude Control)

The attitude control systems have the following functions:

- provide stability of the spacecraft and maintain an orientation;

- rotate the vehicle on command into a specific angular position in order to point in a particular direction;

- correct the orbit perturbations due to the magnetic torque and the solar radiation (if the orbit is high).

The satellite attitude has to be corrected regularly about three mutually perpendicular axes, each with two degrees of freedom, giving a total of six degrees of rotational freedom. In order to apply a true torque, it is necessary to use two thrusters being at an equal distance from the center of mass and having equal thrust level and duration. The pointing precision is a specification of the mission.

The torque T of a pair of thrusters of thrust F and separation distance L is applied to provide the spacecraft with a rotational moment of inertia M_a and an angular acceleration a:

$$T = F \times L = M_a \times a \tag{2.5.12}$$

$$I_t \times L = M_a \int a \, dt . \tag{2.5.13}$$

For a homogeneous cylinder, $M_a = \frac{1}{2} M_S r^2$ and for a homogeneous sphere $M_a = \frac{2}{5} M_S r^2$.

If the angular acceleration is constant over the time period, the satellite will move around its axis of mass at an angular speed $\omega = a \times t$ and through a displacement angle given by

$$\theta = \frac{1}{2} a \times t^2 . \tag{2.5.14}$$

Hence the propellant mass needed to increment the angular velocity of a spacecraft subject to a total angular torque perturbation T_p is equal to

$$m_p = \frac{T_p \times t_c}{g_0 I_{sp} \times L} \tag{2.5.15}$$

where t_c is the duration of one cycle of rotation.

The minimum impulse I_m defines the precision of the angular displacement of the satellite, θ_0, as follows

$$(\dot{\theta})_0 = \frac{1}{2} \frac{L \times I_m}{M_a}$$

and thus, from Equations (2.5.13) and (2.5.14)

$$\theta_0 = \frac{1}{4} \frac{L \times I_m}{M_a} \times t_c . \tag{2.5.16}$$

2.5.3.2.3 Propulsion Requirements for Small Satellites

As described in previous sections, the determination of the propulsion needs for a micro satellite depends on its total mass, volume, area in contact with the sur-

Table 2.5.2. Table of estimated annual needs for a 50 kg satellite

	Primary ΔV (m/s)	Secondary ΔV (m/s)	ACS (m/s)
Low orbit, <1000 km	<10	1–25	<10
Medium	>100	<1	<1
Geosynchronous	~1500	<5	~5

rounding fluid, and the specification of the mission that must be completed. Nevertheless, micropropulsion needs can be divided into different types of micropropulsion module: (1) *primary* ΔV, for orbit transfer and repositioning, (2) *secondary* ΔV, for station keeping, drag compensation, and (3) *attitude control system (ACS)* for attitude control. Microsatellite secondary ΔV may require small impulse bits to achieve secondary ΔVs similar to those of larger satellites.

The precise requirements of ΔV are not well defined by the space agencies because of the lack of experience in 'micro space field'. One paper discussed the propulsion requirements for very small satellites [9]. First evaluations give the thrust level for micropropulsion ranging from 1 µN to 10 N. We can also find in that 10 kg class satellites may require 0.1–10 mN thrusters for typical on-orbit operations such as attitude control and orbit maintenance [1]. Table 2.5.2 gives some ΔV requirements depending on the type of space mission [10, 11].

Propulsion options are chosen as a function of the mission ΔV requirements.

In order to classify the micropropulsion devices, two main categories of devices depending on thrust force creation can be distinguished [12, 13]: electrical thrusters, referring to electrothermal, electrostatic and electromagnetic devices, and chemical thrusters, which use the chemical energy contained in a stored fuel (liquid, solid, or gaseous).

2.5.3.2.4 Electrical Thrusters

Electric propulsion devices use electric energy from a separate energy source for ejecting propellant mass. The basic subsystems of a typical electrical propulsion system contain (1) a power source, (2) a conversion system that converts this energy into electrical forms at the necessary voltage, frequency, pulse rate, and current level, (3) a system that stores and then delivers the propellant, and (4) one thruster to convert the electrical energy into kinetic energy. Among electric thrusters, we can distinguish the following three fundamental types:

- *Electrothermal thruster:* the propellant is heated electrically and thermodynamically expanded, and the heated gas is accelerated to supersonic speeds through a nozzle. In that case, the thermal energy is transformed into kinetic energy. The 'simplest' electrothermal thrusters are resistojets. In resistojets, the propellant is heated by convective contact with a very hot metal. A description of miniaturized resistojets is presented in Section 2.5.4.1. The arc jet thruster also illustrates this class of propulsion device.

- *Electrostatic thruster:* this provides thrust by accelerating a charged plasma by means of a static electric field. It can only operate under vacuum. For efficiency reasons, charged heavy atoms are used. Electrostatic thrusters can be classified depending on the source of charged particles used, as follows:

  - *Electron bombardment thruster:* a plasma is generated from Xenon propellant by electrical discharges and accelerated in an electrostatic field, creating thus a thrust at the exit. Numerous issues have to be regarded closely for the miniaturization: the wall effect, the presence of a negative charge that could be injurious for the satellite.

  - *Field electrical emission propellant [14] (FEEP) or ion thruster:* very small liquid droplets of propellant are charged by an intense electrical field and then accelerated to produce thrust force. An example of miniaturized ion thruster is presented in Section 2.5.4.2.

- *Electromagnetic thruster:* provides thrust by accelerating a charged plasma by means of an electromagnetic field. There is a wide variety of electromagnetic thrusters. Most of them use plasma as part of the current-carrying electrical circuit and all obtain the thrust from magnetic energy. Two types of thrusters have been investigated for miniaturization: the pulsed plasma thruster (PPT) and the Hall thruster (HT). The former operates in the transient mode and the latter in the steady state. Investigations of the miniaturization of the HT seem to be unfruitful because a minimum size greater than 1 cm^2 is required to produce thrust by the Hall effect [15]. In the case of the PPT, a propellant (commonly Teflon) is loaded in a chamber, and a capacitor applies an electric arc at the surface of the propellant, vaporizes several molecular layers and creates a plasma. The electromagnetic field accelerates the plasma and generates a low thrust. The negative and positive species, which exist in the plasma in equal quantities, are accelerated together in the thruster and give a neutral exhaust beam. This is important in order to prevent charge accumulation in the satellite. One example of development of a micro PPT is described in Section 2.5.4.6.

2.5.3.2.5 Chemical Thrusters

Chemical propulsion devices use the chemical energy stored in a fuel substance and convert it into thrust by accelerating the ejected gas stream through a nozzle. The fuel substance is either liquid, solid, or gaseous.

Chemical propulsion systems are made of (1) a propellant tank that can be pressurized, (2) an ignition device that could be an electrical or mechanical part, and (3) a converging–diverging part to accelerate the gas stream. Some require intermediate parts such as valves, pumps, mixing parts, or feeding systems.

Among the chemical propulsion systems investigated for miniaturization, cold gas propulsion systems appear to be relatively simple and relatively cheap. They present good, reproducible performance and can provide very small and precise impulse forces. An example of development is given in Section 2.5.4.3.

The hydrazine liquid monopropellant thruster has proven its capability for large spacecraft. Owing to their high I_{sp} (200–300 s), hydrazine thrusters are good candidates for primary ΔV propulsion systems. Because of the liquid storage, these thrusters require valves and could experience functioning failure. Today efforts are needed to achieve the miniaturization of such a system; important issues are the material compatibility and propellant safety and also the power required to pre-heat the system.

Adding another fuel to the system (hydrazine/nitrogen tetroxide) improves the performance of the systems and increases the specific impulse (~ 350 s). These bi-propellant engines are thus suitable for primary ΔV application. Downscaling these systems implies a technological issue in obtaining good propellant mixing within the small combustion chamber.

Another chemical miniaturized option is the solid propellant thruster approach, which is simple, reliable, and low cost and can give relatively correct specific impulses (100–200 s). Solid propellant thrusters are interesting for the generation of short impulses. The major drawback is the lack of restartability that implies associating thrusters in addressable arrays. Two examples of solid propellant micro thruster arrays development are given in Section 2.5.4.5.

2.5.3.3 Conclusions

The choice of a particular option (electrical or chemical) for the propulsion device is dictated primarily by the specific mission objectives. Then the selection of the propulsion systems for a specific mission is made after an analysis of the available propulsion system characteristics (F, I_{sp}, I_t, I_m) and the mission requirements (altitude, duration, total mass, etc.). Many technological developments have been made and many micropropulsion devices have been demonstrated, characterized with respect to their capabilities, limits, advantages, drawbacks, and possible applicability.

In conclusion, electric propulsion systems offer high I_{sp} (~ 500 s for the arcjet and $\gg 1000$ s for others). For most of them the performance is directly linked to the input power, which is a limitation for their miniaturization. From the evaluation made of chemical propulsion concepts, it appears that two options are complementary and can be miniaturized fairly easily: (1) the solid propellant concept, which gives a good balance between the desired performance (high thrust and secure environment), the level of complexity, and low cost, and (2) the cold gas thruster, which permits very accurate attitude control by delivering very small and precise thrust for long durations.

2.5.4   Emerging Technology in Micropropulsion – Examples of Some Current Developments

In the last decade, substantial technological work has been performed for the miniaturization of thrusters using microsystem technology. The main motivation to explore microsystem technology in micropropulsion is guided by the overall strategy of system miniaturization and cost saving. The increasing interest in microsystem technology is related once again to the gain in mass, cost, performance and reliability that this technology could offer and also the compatibility it offers with electronic technology permitting the best integration.

Among the propulsion systems under investigation for miniaturization and using microsystem technology or hybrid technology are the cold gas thruster [16], subliming solid thruster [17], solid propellant thruster [18, 19], vaporizing liquid thruster or resistojet [20, 21], micro ion thruster [22, 23], and pulsed plasma thruster [24].

2.5.4.1   Vaporizing Liquid Thruster (VLT) (or Microresistojet)

The VLT operates by vaporizing a liquid propellant (water and hydrazine are the most commonly used) inside a micromachined heater assembly. The propellant enters the thruster chip assembly through an etched hole of a few hundred μm^2 into the vaporization chamber (see Figure 2.5.3). Deposited resistors heat the fluid and vaporize it. The fluid flows along a channel micromachined into the wafer. The propellant vapor exits the chip assembly through a micromachined nozzle. The throat diameter is about 50×50 μm. The number of wafers stacked together to realize the assembly can be two or three.

The advantages of VLT are its very small size and so weight, its scalability to very low proportions and the precise thrust impulse. It does not contain any moving parts, which makes it reliable. The VLT thrust is several millinewtons with an input power of a few watts.

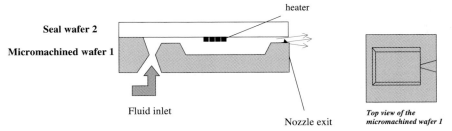

Figure 2.5.3. Schematic view of a vaporizing liquid micro thruster having the outlet nozzle on the side.

This option is well adapted for the attitude control of the nano and micro spacecraft. The drawbacks are the leakages that could occur during storage and the level of input power needed to vaporize the liquid propellant.

2.5.4.2   Micro Ion Thruster (μIT)

The μIT is a miniaturization of a conventional ion engine in which a low-pressure gas discharge is created through bombardment by electrons generated by a cathode. Ions are extracted from the gas discharge and are accelerated electrostatically to high velocities (about 30000 m/s) in a set of accelerator grids being at a voltage of 1 kV (see Figure 2.5.4). Currently, a 3 cm diameter micro ion engine is under investigation at Jet Propulsion Laboratory (JPL) for micro spacecraft. Micro ion engine development has not yet been performed. Many technical difficulties have been encountered in miniaturizing the device subsystems such as the cathodes, the neutralizers, and the grids [22]. Each of these subsystems requires extensive technical investigations.

Micro ion thruster concepts may be proposed for primary propulsion devices for micro and nano spacecraft, because of their very high specific impulse (~ 300 s). The high level of the specific impulse results in a decrease in the propellant consumption to achieve a high ΔV requirement. μIT could also find application in the field of attitude control for large spacecraft.

2.5.4.3   Micro Cold Gas Thruster (CGT)

The CGT concept consists of ejecting gas through a micro nozzle at high speed in order to create force. The Angström Space Laboratory of Uppsala University proposes a cold gas system solution consisting of a stack of three micromachined subsystems: (1) the nozzle unit with heat exchangers consisting of four nozzles generating supersonic flow; (2) the valve unit consisting of four proportional piezoelectrically actuated valves; and (3) the filter unit forming a stack of three silicon wafers. The assembly is connected to a reservoir as shown in Figure 2.5.5.

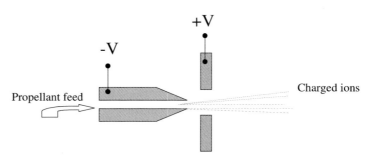

Figure 2.5.4. Schematic view of a micro ion thruster.

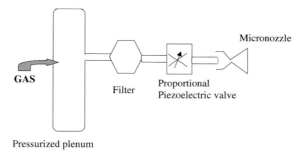

Figure 2.5.5. Schematic view of the gas system for a cold gas thruster..

The CGT can deliver from 0.1 to 10 mN and is well adapted for fine attitude control. It has two main drawbacks: the extremely low I_{sp} (~ 60 s) and the high leakage possibility, reducing the reliability for long mission durations.

The CGT is suitable for low ΔV applications on micro spacecraft and for attitude control of small satellites.

2.5.4.4   Subliming Solid Micro Propulsion (SSP)

The SSP concept is not new and substantial development work was performed with this kind of thruster in the 1960s. It is based on the decomposition of solid propellant having a high sublimation pressure. Upon heating, the pressure in the propellant tank increases and the vapor is exhausted through a valve and nozzle to produce thrust force. A MEMS-based subliming solid thruster [2, 17] consists of two layers, one in silicon and one in Pyrex (see Figure 2.5.6): (1) in the silicon is micromachined a chamber (~ 6 mm diameter and 550 µm height) and a

Figure 2.5.6. Schematic view of a subliming thruster.

nozzle (throat diameter 50 µm), with a nozzle expansion angle of 27.7°; and (2) the Pyrex part on top of the silicon chip contains a hole driving the flow.

The propellant used is a heterocyclic nitrogen compound. The propellant vapor enters the chip from the tank towards the nozzle hole. On its way, the propellant passes through a micromachined filter that prevents solid particles from leaving the tank.

Assuming a nozzle area ratio of 100, the theoretical performance for the compound as rocket propellant is ~250 s. The application of this type of technology could be in the attitude control of small satellites.

2.5.4.5   Solid Propellant Thruster (SPT)

The concept is based on the high rate of combustion of a single propellant stored in a thrust chamber. The gas generated by the combustion of the propellant is accelerated in a nozzle, thus delivering a thrust. Because they are one shot, SPTs are fabricated in arrays. Two different approaches were followed: (1) one initiated at TRW, a digital thruster [19], and (2) one initiated at LAAS-CNRS [18].

The TRW digital thruster approach consists of a three-layer structure (silicon and glass) containing a micro resistor wafer, a thrust chamber, and a rupture diaphragm (see Figure 2.5.7). An explosive is loaded into each individual sealed chamber, which is made of photoetchable glass. When the resistor heats, the propellant ignites, the pressure in the chamber rises and the diaphragm breaks. A thrust impulse is generated. Typical dimensions are a few hundred μm^2 for the exhaust holes (made by KOH etching and covered with an SiN diaphragm), around 700 µm for the chamber diameter and 1.5 mm for the chamber length. Initial characterization made with lead styphnate as the material gave a performance of 0.1 mN for the thrust.

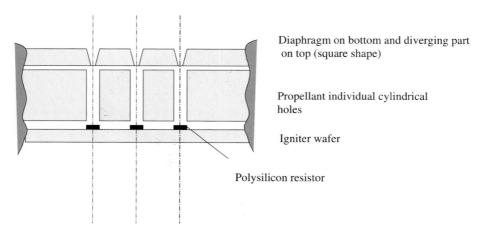

Diaphragm on bottom and diverging part on top (square shape)

Propellant individual cylindrical holes

Igniter wafer

Polysilicon resistor

Figure 2.5.7. Schematic view of the digital thruster device.

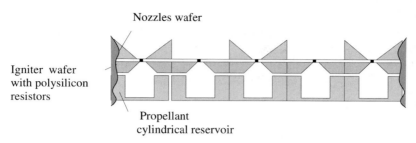

Igniter wafer
with polysilicon
resistors

Nozzles wafer

Propellant
cylindrical reservoir

Figure 2.5.8. Schematic view of the solid propellant thruster structure.

The objective of this research is to have thrusters for high-accuracy station keeping and attitude control of very small spacecraft.

The SPT developed at LAAS-CNRS is based on the gas issuing from the deflagration of a glycidyl azide polymer-based propellant. The single thruster consists of a stack of three parts of silicon (see Figure 2.5.8): (1) a silicon micromachined igniter with a polysilicon resistor patterned on to a very thin dielectric membrane. (2) a propellant chamber; and (3) a diverging part wafer can be added on top of the structure if necessary. The main difference from the TRW digital thruster option is that the propellant is heated at the throat and not at the bottom of the thrust chamber, thus allowing perfectly controlled combustion. The micromachined silicon igniter membrane has a double function: as a heat source that initiates the combustion and as the throat of the nozzle. For a structure having a throat diameter of 110 μm, a chamber diameter of 850 μm, and a chamber length of 1 mm, the expected force is around 5 mN. The total impulse is about 1.5 mN s.

The short duration of the thrust impulse added to the fact that the thruster is not restartable make the SPT suitable and well-adapted for short-duration missions to achieve a low ΔV requirement and station keeping of small satellites.

2.5.4.6   Micro Pulsed Plasma Thruster (μPPT)

The μPPT is an electromagnetic thruster and its concept has already been described above. The μPPT [24] consists of a bar of Teflon propellant pressed between two electrodes with a spring. The spring ensures that the propellant remains in contact during consumption. The capacitor is charged to a voltage of 500–2000 V. Then a discharge is ignited and all the energy stored in the capacitor (5–50 J) powers a high-current/short-duration plasma discharge. The discharge ionizes a small amount of the propellant and accelerates it to high velocities (well-known Lorentz law) of up to 4000 m/s. The total mass of the μPPT is 3.8 kg, which is not a miniature system owing mainly to the capacitor. The key issue for the μPPT now is to demonstrate its capability for miniaturization. The resulting I_{sp} (~ 500–1000 s) makes this technology well suited for primary ΔV for nano satellites. The principle of a PPT is illustrated in Figure 2.5.9.

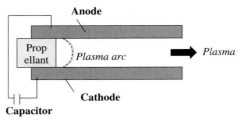

Figure 2.5.9. Pulsed plasma thruster concept.

2.5.4.7   Summary and Table of Comparison

Cold gas thrusters and subliming solid thrusters are relatively simple and inexpensive. The major limitation of both of these thrusters is the low specific impulse. Cold gas thrusters also have the disadvantage of a high leak rate. Miniaturized vaporizing liquid thrusters have higher I_{sp} than cold gas thrusters and solid subliming thrusters but are very complex and high powered. Micro ion thrusters are complex and also require high input power but present very interesting I_{sp}. Micro solid propellant thrusters are simple, inexpensive and have an acceptable performance but cannot be reused. Their good adaptability makes the pulsed plasma thruster a good

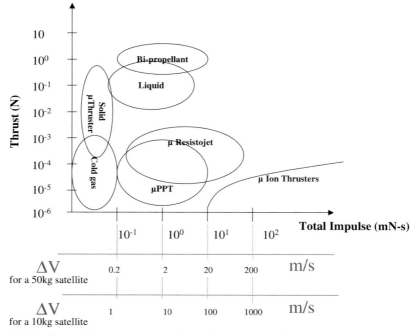

Figure 2.5.10. Chart of operating range for micro thruster concepts.

Table 2.5.3. Table of comparison of the micropropulsion systems characteristics and performance

Micro thruster type	Thrust range (mN)	Specific impulse (s)	Total impulse (mN s)	I_{min} (µN s)	Thrust duration	Power (W)	Specific power* (mN/W)	Applicability for micro and nano spacecraft
Resistojet (hydrazine)	50	300			Long	100–500	2–10	ACS
Resistojet (water)	0.05–0.5	100			Long	5–10	50	ACS
µPPT	0.15	500	0.1–1	70	Long	10	60–70	Prim. ΔV
µIon µFEEP	0.001–1	3000	15000		Very long	10–20	10–20	Prim. ΔV ACS
Arcjet	150	465			Long	1400	10	Prim. ΔV
Liquid thruster (hydrazine)	10–200	200	0.1–2	10	Medium	10	20	Prim. ΔV
Cold gas	10^{-3}–10	65	0.1	10	Long	10	2	Sec. ΔV ACS
Solid propellant – composite propellant	1–200	~100	0.1–1		Short impulse	1	0.005	Sec. ΔV
Solid propellant – lead styphnate	0.1		0.1	50	Extremely short impulse	150	500	Sec. ΔV ACS
Bi-propellant	2000	265	30		Medium	18	0.009	Prim. ΔV

* Thrust force over the input power.

candidate for low primary ΔV, secondary ΔV and ACS. Each concept and option presents its own advantages and capabilities and has its own limitations.

Table 2.5.3 summarize the characteristics and performance of each option.

In Tables 2.5.4 and 2.5.5 are reported the advantages and drawbacks of each micro thruster option.

The chart of Figure 2.5.10 is a two-dimensional classification of the micro thruster operating performance: thrust force (N) over the total impulse (mN s) space.

The chart in Figure 2.5.11 is a two-dimensional classification of the micro thruster operating performance: thrust force (N) over the thruster efficiency (mN/W) space.

Table 2.5.4. Advantages and drawbacks of electrical propulsion systems

Micro thruster type	Advantages	Drawbacks
Resistojet	Simple device Easy to control Can use inert propellant such as water Low cost	Lowest I_{sp} Heat loss
Ion thruster	High specific impulse, High efficiency Use inert propellant (Xe)	Complex power conditioning High voltage Low thrust per unit area Heavy power supply
Pulsed plasma thruster	Simple device Low power Solid propellant (Teflon) so no leakage possible	Low thrust Teflon reaction products are toxic, corrosive
Arcjet	Direct heating of gas Low voltage Relatively simple device Inert propellant	Low efficiency Erosion at high power Low I_{sp} Heat loss More complex power conditioning

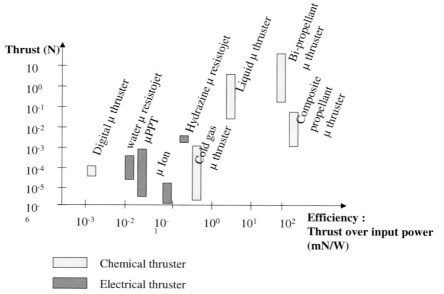

Figure 2.5.11. Chart of thruster efficiency (thrust force over input power) for the different concepts currently developed.

Table 2.5.5. Advantages and drawbacks of chemical propulsion systems

Micro thruster type	Advantages	Drawbacks
Liquid propellant	Often high specific impulse Can be stopped and restarted Ready to operate quickly	Relatively complex design Large number of moving parts and components Possible leaks during the storage Tanks need to be pressurized with a separate pressurization system Difficult to control combustion instability Usually required more volume Most liquid propellant are corrosive, toxic, and hazardous
Solid propellant	Simple (no moving parts) Easy to operate and quick in operation Stable with time so storable for many years Usually allows more compact package, interesting for small satellites Some propellants has non-toxic combustion gases Thrust determination devices permit some control over total impulse, so are adaptable to mission requirements by choosing the best propellant and the appropriate geometric thruster features Can be designed to operate in a very stable regime of combustion	Some propellants can deteriorate during long storage Under certain conditions, some propellants may detonate Many propellants contain metal particles and the exhaust is smoky Performance will vary with ambient temperature Cannot be tested and verified before operating

As already mentioned, the criteria for the selection of a particular spacecraft propulsion system depend on the mission and applications. However, some criteria are applicable at all the missions:

- *System performance:* the thrust-time profile and the number of impulses available are selected to optimize the mission.

- *Safety and reliability:* the risk of failure or damage needs to be evaluated and minimized. Inadvertent energy input to the propulsion system may not result in a detonation. Adequate storage time of components and propellant may be demonstrated.

- *Controllability:* the combustion process must be stable. The time response to command must be acceptable. The reproducibility of the thrust-time profile must be the optimum.

- *Volume:* the propulsion system must fit the volume constraints of the space-craft: the smallest is always the best.

- *Operability:* simple to operate is the best.

- *Cost:* easy to manufacture and production must be low cost.

- *Power constraint:* as the power available decreases with the miniaturization of the spacecraft, the micropropulsion systems must have low power consumption.

The solid propellant thrusters respond to several criteria: minimization of the power consumption, low fabrication cost, simple device minimizing the risk of failure, easy operability and safety requirements because of using a non-vulnerable propellant, and possibility of making very short and precise impulses. For low ΔV, fitting the best to its performance range, solid propellant thrusters may be interesting compared with other options. A detailed presentation of solid propellant thrusters is given to illustrate the interest in microsystem technology in the micropropulsion field.

2.5.5   A New Millinewton-scale Micropropulsion Device: the Solid Propellant Micro Thruster Array

2.5.5.1   Micro Fabrication of the Device

The micro thruster prototypes we present in this paper have the features and dimensions shown in Figure 2.5.12.

Two kinds of devices have been designed and fabricated, one having a throat diameter of 108 µm and a chamber diameter of 850 µm, and the other having a throat diameter of 250 µm and a chamber diameter of 1 mm. All other dimensions are kept constant.

2.5.5.1.1 Thermal Igniter Fabrication

A 4 inch (100)-oriented silicon wafer is thermally oxidized. The wafers are then coated with silicon-rich LPCVD (low-pressure chemical vapor deposition) nitride. The resulting thickness is 0.7 µm. In a third step, a layer of 0.5 µm of poly-silicon is deposited by LPCVD at 605 °C and patterned using a reactive ion etch-

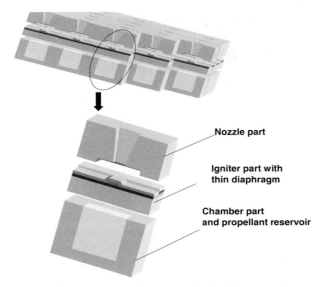

Figure 2.5.12. Scheme of the geometric features of a single micro thruster.

ing (RIE) plasma of CF_4 and O_2 in order to design the resistor. Then, the electrical pads and electrical supply lines are realized in gold. A square window is opened in the dielectric layer on the rear side of the wafer by conventional photolithography and RIE etching in a CF_4 and O_2 plasma. The silicon is then anisotropically etched away with KOH. The fabrication yield of the structure with an SiO_2/SiN_x membrane 0.7 µm thick is very close to 100%. A scanning electron microscope (SEM) photograph of one microheater array and an SEM photograph of a single igniter with the polysilicon resistor are shown in Figures 2.5.13–2.5.15.

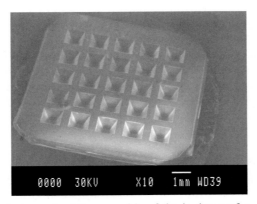

Figure 2.5.13. SEM photograph of the rear side of the igniter wafer.

Figure 2.5.14. SEM photograph of the electrical connections of the igniter wafer.

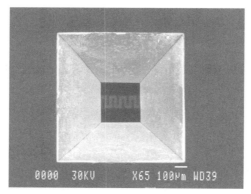

Figure 2.5.15. SEM photograph of a single igniter (view from the rear side).

2.5.5.1.2 Fabrication of the Silicon Chambers

The silicon chambers are realized by deep reactive ion etching (DRIE). In Figure 2.5.16, part of a propellant chamber array of dimension 9.5×9.5 mm is shown. In total, it contains 25 square holes with a depth of 525 µm obtained by DRIE.

An alternative material for the fabrication of chamber is a ceramic. Ceramic is very interesting because of its low thermal conductivity compared with silicon. So far, we have fabricated ceramic chambers of 1 mm of diameter by conventional drilling in MACOR®. For further development, LAAS has chosen to orient the investigation towards ceramic injection molding to realize a micro chamber.

2.5.5.1.3 Filling and Assembling

Filling the micro chamber with viscous material was a key issue for the success of such a device. Indeed, cavity-free filling of through holes or blind holes with a pasty

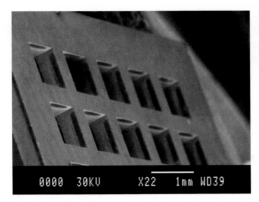

Figure 2.5.16. SEM photograph of propellant chamber array.

product featuring an important aspect ratio (depth/diameter) is always critical. The air trapped in the blind hole or in the cavity is a limit to the filling and a low-viscosity products require a high pressure in order to be injected into a small aperture without damaging the thin membrane. Previously some vacuum chamber injection machines have been developed and tested. The equipment consists of a vacuum chamber in which the injection system is installed. The drawbacks of this kind of device are their cost and their complexity compared with their low efficiency.

The present innovation (patented by Novatec) brings a new and original solution to the problem of filling holes without the drawbacks of the previous systems, and can be fitted on existing stencil printers. The technology consists of a specific transfer head which slides across the transfer stencils or directly on top of the substrate to be filled. Compared with existing systems, this innovation shows very high efficiency and a high reliability level for processing in addition to its low equipment cost. The technology is particularly well suited to plug through holes or blind holes on a substrate with paste in a collective manner. Figure 2.5.17 shows a schematic view of the principle of the process and the associated equipment.

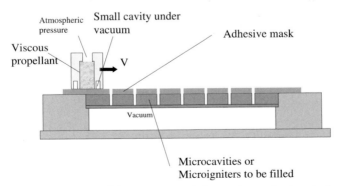

Figure 2.5.17. Schematic view of the process and the equipment set up to load the propellant inside micro cavities.

Each part of the device (the igniter and the reservoir) is first filled with the propellant and then assembled with epoxy glue, enabling the entire structure to support the combustion pressure inside the combustion chamber. We used EPO TEK H70E glue cured at 60 °C for 15 h. The experimental results demonstrated that the gluing can withstand a pressure of 30 bar in the thrust chamber.

2.5.5.2 Solid Propellant Thruster Performance Evaluation

2.5.5.2.1 Model Description

The thermodynamic relations for the processes occurring inside the thruster chamber and nozzle give the mathematical tools needed to calculate the performance. They permit the prediction of the operational performances of thrusters depending on their shape and size. The well-known equations governing a fluid flow are the conservation of mass, momentum, and energy (the full Navier-Stokes equations for a compressible fluid [25]):

$$\frac{\partial \rho}{\partial t} + \nabla \cdot (\rho v) = 0 \tag{2.5.17}$$

$$\frac{\partial (\rho v)}{\partial t} + \nabla \cdot (\rho v \times v + pI) = \rho f + \nabla \cdot \tau \tag{2.5.18}$$

$$\frac{\partial (\rho e)}{\partial t} + \nabla \cdot (\rho h v) = \rho f \cdot v + \nabla \cdot (v \tau) + \nabla \cdot Q \tag{2.5.19}$$

where ρ is density, v is velocity, p is gas pressure, f is an external volume force per unit mass, τ is the viscous stress tensor, $e = e_{int} + v^2/2$ is the specific energy, e_{int} is the internal energy, $h = e + p/\rho$ is the specific enthalpy, and Q is the heat transfer rate.

In order to solve this system of differential equations, it is necessary to specify the equation of the state of the gas and its energetic equation.

In the case of a perfect gas, the gas law and the energy are given by

$$p = \rho R' T \tag{2.5.20}$$

$$e = \int c_v dT \tag{2.5.21}$$

where T is the temperature, c_v is the specific heat capacity and R' is the gas constant.

2.5.5.2.2 Main Assumptions

- Pressure and temperature are considered to be homogeneous in the burning chamber.

- Burning gas is a perfect gas.

- Flow in the diverging and in the throat is assumed to be isentropic, one-dimensional, and quasi-static.

- The pressure variation on the burn rate law is neglected.

2.5.5.2.3 Model Computation

A lumped parameter approach was chosen for our application [26–29]. This approach consists in building several local models and integrating them into a global compact model with SIMULINK, a tool of Matlab. Propellant characteristics are given in Table 2.5.6.

2.5.5.2.4 Performance Evaluation for a Single Thruster – Computational Results

Results of thrust force impulse have been reported to illustrate the performances of solid propellant micro thrusters: the thrust impulses for a thrusters featuring cross-sectional area ratios $A_c/A_t = 60$ ($\phi_t = 108$ μm; $\phi_c = 850$ μm) and $A_c/A_t = 16$ ($\phi_t = 250$ μm; $\phi_c = 1000$ μm), respectively (ϕ_t = throat diameter, ϕ_c = chamber diameter).

The calculations were performed for two conditions: when the flow leaves the thruster at atmospheric pressure (see Figure 2.5.18) and when it leaves at a surrounding pressure of 1 mbar (see Figure 2.5.20).

Table 2.5.6. Characteristics of the propellant used in micro solid propellant thrusters

Propellant property	Composite propellant
Composition (main compound)	Oxidizer (NH_4ClO_4), fuel, binder
C (J/g)	2900
$V = aP^n + b$	
a (m/s Pa)	7.76×10^{-5}
n	0.29
b (m/s)	1×10^{-3}
$\gamma = C_p/C_v$	1.3
ρ_p (kg/m^3)	1500
C_p	1885

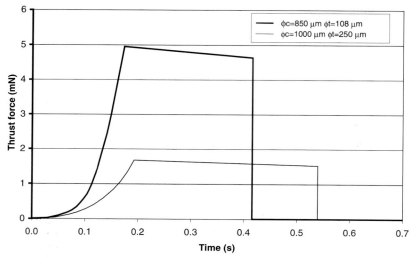

Figure 2.5.18. Thrust force as a function of time (external pressure at atmospheric).

At atmospheric pressure, a thruster featuring a chamber-to-throat ratio (A_c/A_t) of 60 delivers roughly 4.8 mN over 250 ms. A thruster featuring a chamber-to-throat ratio of 16 delivers less than 1.5 mN during 350 ms. This low thrust force is due to the subsonic flow in the throat.

Under vacuum, the thruster featuring a chamber-to-throat ratio of 60 delivers roughly 5.8 mN over 250 ms. For $A_c/A_t=60$ the thrust force is higher than when the external pressure is atmospheric. This is explained by the added force due to atmospheric pressure which is equal to 0.9 mN. The thrust generated by the gas flow is reduced by this force. The thruster featuring a chamber-to-throat ratio of 16 delivers 5 mN during 350 ms. This thrust force increase occurs because of sonic flow in the throat and to the atmospheric force disappearance.

2.5.5.3   Summary of Solid Propellant Micro Thruster Performance

To assess the capability of the solid propellant thrusters for space applications and in order to have figures to evaluate this technology for particular mission application, we have reported in Table 2.5.7 different sizes of thrusters and their corresponding performances.

The graphs in Figures 2.5.20 and 2.5.21 give the thrust range and the total impulse range accessible with solid propellant technology.

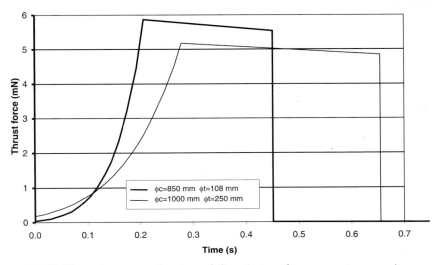

Figure 2.5.19. Thrust force as a function of time (external pressure at vacuum).

Table 2.5.7. Summary of solid propellant characteristics and performance at atmospheric pressure

Chamber diameter ϕ_c (μm)	Throat diameter ϕ_t (μm)	Chamber-to-throat section ratio, A_c/A_t	Thruster length, L (mm)	Chamber pressure, bar	F (mN)	I_t (mN s)	I_{sp} (s)
300	50	36	1.5	2.5	0.3	0.15	90
500	83	36	1.5	2.5	1	0.4	102
850	141	36	1.5	2.5	3	1	109
850	85	100	1.5	9.5	6	1.4	153
850	55	238	1.5	31	9	1.6	167
1000	1667	36	1.5	2.5	4.3	1.4	110
1000	100	100	1.5	9.5	8.5	1.8	158
1000	65	236	1.5	31	12	2	163
1500	250	36	1.5	2.5	10	2.5	113
1500	150	100	1.5	9.5	20	3.4	147
1500	100	225	1.5	30	28	4	179

Thrust force

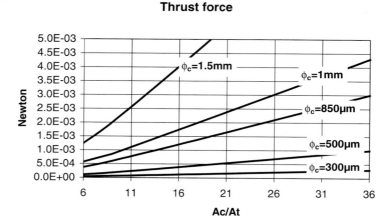

Figure 2.5.20. Thrust force as a function of the chamber-to-throat section ratio for five different chamber sections.

Impulse bit force

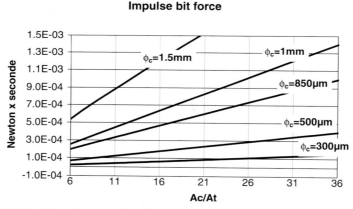

Figure 2.5.21. Total impulse as a function of the chamber-to-throat section ratio for five different chamber sections.

2.5.5.4   Conclusions

Despite the fact that the use of solid propellants for thrust generation is not an original contribution, it appears to be inevitable for downscaling of thrusters and rockets. The fabrication process of the proposed solid propellant micro thrusters is based on conventional microsystems technology and the assembly is very simple to perform. The feasibility of using low thermal conductivity materials such as ceramic for the propellant tank leads to miniaturization of heat transfer between the thrusters. This opens up the way for thruster downscaling of the devices. The dependence of the resulting thrust on dimensional and geometric parameters and on propellant characteristics implies the possibility of fabricating micro thrusters with a total impulse ranging from 0.1 to 2.5 mN s.

The model of combustion behavior briefly presented above requires short calculation times (a few minutes), is applicable to the subsonic and supersonic regimes and is easy to adapt to different materials and shapes of thrusters, thus permitting optimization of micro thruster arrays.

2.5.6 Conclusions – Microsystems in the Micropropulsion Field: Advantages, Limits and Future Development Needs

As the literature illustrates [4, 10, 30], the technological effort in micropropulsion is mainly devoted to microsystem technology. The reasons for exploring microsystem technology in the fabrication of micropropulsion systems are the gain in size, the increase in performance, and the possibility of increasing integration.

The first reason, gain of size, is the best known and discussed. Using microsystem technology, very small devices can be fabricated with very precise shapes (1 µm). In particular, for application in the future generation of satellites, very light-weight and small devices are required. Microsystem technology would enable their fabrication.

The second reason, increase in performance, is very important. By exploiting the capability to fabricate extremely small feature sizes with very tight tolerances, it is possible to increase the precision of the thruster performance and to reduce the level of thrust and access to micro thrust impulse.

Finally, microsystem technology will enable us to reach a much higher level of integration of the propulsion components and of the micropropulsion components and their drivers and power conditioning electronics. These modules will feature minimal external interfaces and electrical connections and could be directly attached to the propellant tank. Furthermore, the whole satellite subsystems could be designed and built using microsystem technologies. Miniature sensors, data acquisition, photovoltaic arrays, communication antenna, etc., could all be fabricated on silicon.

There are still technological challenges to be overcome in the development of the individual components:

- the reduction of the dimensional features to reach very precise and short impulses;

- the interfacing, interconnection, and packaging of the systems to have a light and fully integrated system;

- the management in a very small volume of high temperature and high pressure that could lead to mechanical failures to fabricate a sure and reliable system;

- the limits of silicon and the introduction of the most appropriate material for different interfaces.

Dimensions. These are of importance in the development and design of micropropulsion devices. An increased surface-to-volume ratio of the combustion chamber (in the case of a chemical thruster) is a challenge as it leads to heat losses from combusting propellant, affecting the thruster performance and efficiency. Viscous flow losses in small-diameter nozzles may reduce the available specific impulse. Most existing ballistic models do not take into consideration the disturbance caused by the high surface-to-volume ratio. The newest and well-adapted models based on solid-state and fluidic theory are required [31].

Packaging. Most of currently available microsystem technology uses silicon and has developed silicon-based processes. In the case where a ceramic or glass wafer needs to be interfaced with micromachined propulsion components, and also when packaging of delicate micromachined devices is required, novel interfacing and bonding techniques need to be investigated. Since most microfabrication is performed in silicon, as the basic material, glass-to-silicon or ceramic-to-silicon bonding techniques may be required to assemble a full micropropulsion module. Low-temperature glass-to-silicon bonding (below 200 °C) should be available. Further development of ceramic-to-silicon gluing or bonding techniques would be advantageous and need to be investigated.

Silicon vs Non-silicon Fabrication. As already mentioned, most microfabrication techniques use silicon as the base material. The properties of silicon limit the feasibility of the propulsion concept. Most concepts studied create harsh environmental conditions in which all material properties are used at their extremes (especially temperature and mechanical strain). For example, in the case of propellant combustion, the silicon thrust chamber can reach several hundred kelvin and several tens of bar pressure.

Silicon is also a perfect heat conductor with a heat conductance of ~ 140 W/mK. While this could be an advantage for the microelectronics industry in helping to cool the chip, it is a serious drawback for propulsion applications where heat needs to be conserved and constrained to the propellant. Heat losses from the thruster chip reduce the thruster efficiency, and could also damage the structure. Oxidizing the silicon in surface is not a sufficient solution for the reduction of the heat losses. Other design solutions or material types may be used to reduce the heat conduction efficiently.

In conclusion, the development of silicon technology up to now, especially in microelectronics and micromechanics, makes it the base material for microsystems developments. Microsystems technologies have sufficient assets to serve as a basis for micropropulsion developments and have the merit of being adaptive by offering many technical options for fabrication, assembling and integration. In the author's opinion, future developments may also explore hybrid approaches and may explore, in parallel with Si, the newest materials that will suit the micropropulsion requirements best.

2.5.7 Acknowledgments

The work described in this paper was funded by the European Commission through the IST program (MICROPYROS project) and by the National Center for Spatial Studies and Development (CNES). I thank the French groups Lacroix and Novatec for their collaboration in GAP propellant formulation and solid propellant thruster filling, respectively. I also thank all the people at LAAS-CNRS involved in micropropulsion activities: D. Estève, S. Orieux, B. Larangot, A. Berthold, and the LAAS microfabrication team.

2.5.8 References

[1] Helvejian, H., *Microengineering Aerospace Systems;* AIAA, 1999.
[2] Mueller, J., Muller, L., George, T., *Subliming Solid Micro-Thruster for Microspacecraft*, New Technology Report NPO-19926/9525.
[3] Youngner, D.W., Lu, S.T., Choueiri, E., Neider, J.B., Black, R.E., Graham, K.J., Fahey, D., Lucus, R., Zhu, X., presented at the 14th Annual/USU Conference on Small Satellites, 2000.
[4] Mueller, J., Chakraborty, I., Vargo, S., Marrese, C., White, V., Bame, D., Reinicke, R., Holzinger, J., presented at the IEEE Aerospace Conference, Big Sky, Montana, March 18–25, 1999.
[5] Ye, X., Tang, F., Ding, H., Zhou, Z., *Sens. Actuators A* **89** (2001) 159–165
[6] Murkerjee, E.V., Wallace, A.P., Yan, K.Y., Howard, D.W., Smith, R.L., Collins, S.D., *Sens. Actuators A* **83** (2000) 231–236.
[7] London, A.P., Ayon, A.A., Epstein, A.H., Spearing, S.M., Harrison, T., Peles, Y., Kerrebrock, J.L., *Sens. Actuator A* **2997** (2001) 1–7.
[8] Sutton, G.P., *Rocket Propulsion Elements*, 6th edn., New York: Wiley, 1992.
[9] de Groot, W.A., *Propulsion Options for Primary Thrust and Attitude Control of Microspacecraft*, NASA/CR-1998-206608.
[10] Mueller, J., Leifer, S.D., Muller, L., George, T., *Design, Analysis and Fabrication of a Vaporizing Liquid Micro-Thruster*, AIAA Paper 97-3054, Seattle, 1997.
[11] LAAS Report No. 96304, 1996.
[12] Jonson, S.W., presented at the 30th AIAA/ASME/SAE/ASEE Joint Propulsion Conference, June 27–29, 1994, Indianapolis, IN.
[13] Mueller, J., presented at the 33rd AIAA/ASME/SAE/ASEE Joint Propulsion Conference, July 6–9, 1997, Seattle, WA.
[14] Marcuccio, S., Genovese, A., Andrenucci, M., *Journal of Propulsion and Power,* **14** (1998) 774–781

[15] Khayms, V., Martinez-Sanchez, M., *Design of a Miniaturized Hall Thruster for Microsatellites*, AIAA Report, 1996.

[16] Kölher, J., Simu, U., Bejhed, J., Kratz, H., Jonsson, K., Nguyen, H., Bruhn, F., Hedlund, C., Lindberg, U., Hjort, K., Stenmark, L., presented at the 11th International Conference on Solid-State Sensors and Actuators, Munich, June 10–14, 2001.

[17] de Groot, W.A., Reed, B.D., Brenizer, M., presented at the 34th Joint Propulsion Conference, July 13–15, 1998, Cleveland, OH.

[18] Rossi, C., Estève, D., Larangot, B., Orieux, S., *Smart Mater. Struct.* **10** (2001) 1–7.

[19] Lewis, D.H., Janson, S.W., Cohen, R.B., Antonsson, E.K., *Sens. Actuators A* **80** (2000) 143–154.

[20] Ketsdever, A.D., Wadsworth, D.C., Vargo, S.E., Muntz, E.P., *J. Micro Electro Mech. Syst. (1999)*.

[21] Sweeting, M.N., Lawrence, T., Leduc, J., *Proc. Inst. Mech. Eng. Part G* **213** (1999) 223–231.

[22] Brophy, J.R., Polk, J.E., Blandino, J., Mueller, J., *NASA Tech. Briefs* **50** (2) (1996) 36–37.

[23] Gonsalez, J., Saccocia, G., Rohden, H. von, presented at the 23rd International Electric Propulsion Conference, Sept 13–16, 1993, Seattle, WA.

[24] Cassady, R.J., Hoskins, W.A., Campbell, M., Rayburn, C., IEEE, 2000.

[25] Fletcher, C.A.J., *Computational Techniques for Fluid Dynamics,* Berlin: Springer, 1991.

[26] Geider, O., *PhD Thesis*, Institut National des Sciences Appliquées de Toulouse, 1997.

[27] Carrière, P., Ecole Nationale Supérieure de l'Aéronautique et de l'Espace, 1980.

[28] *Solid Rocket Motor Internal Insulation,* NASA SP-8093, 1976.

[29] Ramamurthi, K., Muthunayagam, A.E., presented at the 13th International Symposium on Space Technology and Science, Tokyo, June 28–July 3, 1982.

[30] Stenmark, L., presented at the 11th International Conference on Solid-State Sensors and Actuators, Munich, June 10–14, 2001.

[31] Bayt, R.L., et al., *A Performance Evaluation of MEMS-based Micronozzles*, AIAA Report.

List of Symbols and Abbreviations

Symbol	Designation
a	spacecraft acceleration
A	surface area
A_e	exhaust nozzle area
c_v	specific heat capacity
C_d	drag coefficient
e	specific energy
e_{int}	internal energy
f	external volume force per unit mass

Symbol	Designation
F	thrust force
F_d	drag
g_0	gravitational acceleration
h	specific enthalpy
I_m	minimum impulse
I_{sp}	specific impulse
I_t	total impulse
k	ratio of specific heats (C_p/C_v)
k_d	diffuse coefficient of reflectivity
k_s	specular coefficient of reflectivity
L	distance between spacecraft center of mass and the center of solar pressure
l	distance between two thrusters
$\dot{m}$	mass flow rate
m_p	propellant mass
M	molecular weight
M	satellite/spacecraft mass
M_a	rotational moment of inertia
p	pressure
p	solar radiation pressure
P_1	stagnation condition pressure
P_e	pressure at exit section
Q	heat transfer rate
R	orbital altitude
R	universal gas constant
R'	reduced gas constant
t	time
t_c	duration of one cycle of rotation
T	temperature
T	torque
T_1	stagnation condition temperature
T_p	magnetic torque
T_p	total angular torque perturbation
v	velocity
V	spacecraft flight speed
V_e	exhaust velocity
W	ballistic coefficient
α	angular acceleration
η	propellant mass fraction
ϕ_c	chamber diameter
ϕ_t	throat diameter
φ	angle between incident direction and the normal of the impacted surface

Symbol	Designation
ρ	fluid density
τ	viscous stress tensor
θ	displacement angle
ω	angular speed

Abbreviation	Explanation
ACS	attitude control system
CGT	micro cold gas thruster
DRIE	deep reactive ion etching
FEEP	field electrical emission propellant
HT	Hall thruster
LPCVD	low-pressure chemical vapor deposition
μIT	micro ion thruster
μPPT	micro pulsed plasma thruster
PPT	pulsed plasma thruster
RIE	reactive ion etching
SEM	scanning electron microscope
SPT	solid propellant thruster
SSP	subliming solid micropropulsion
VLT	vaporizing liquid thruster

Index